The Law Doctor 2

Aaron W. Wemple

Life Winks LLC

To all the brave souls who dare to challenge the norms of devices and family law, who navigate the shadows with courage and compassion, and who envision a brighter future for families everywhere. Your resilience, determination, and unwavering belief in justice inspire us all.

"Justice is not a distant ideal but a living, breathing force that must be nurtured within every heart."
— Mahatma Gandhi

Contents

Foreword

In Honor of God and Mahatma Gandhi's Vision

As I stand here amidst the pages of "The Law Doctor 2," I am reminded of the timeless wisdom and unwavering dedication of Mahatma Gandhi, a beacon of light in the history of humanity. Gandhi, with his profound understanding of justice, nonviolence, and human compassion, continues to inspire generations across the globe.

In these pages, you will encounter stories of resilience, innovation, and the relentless pursuit of a more just world—a world where families can thrive, children can dream without fear, and the fabric of society is woven with empathy and understanding.

Gandhi once said, "Be the change that you wish to see in the world." This sentiment echoes through the narratives that unfold in "The Law Doctor 2," where individuals and communities embody the spirit of change, striving to transform the landscape of family law with integrity and compassion.

As we delve into these pages, let us carry with us the timeless teachings of Gandhi and the countless heroes who have championed the cause of justice. May their voices resonate within us, guiding our actions toward a future where justice is not just a concept but a lived reality for all.

With deepest respect and admiration,

Aaron W. Wemple

Preface

In the vast expanse of legal literature, family law stands as a cornerstone of societal structure and human relationships. It is a domain where the intricacies of familial bonds intersect with the cold precision of legal frameworks. "The Law Doctor 2: Delivering juSTEMice to You" emerges as a beacon on ultra-safety in this landscape, shedding light on the nuanced challenges and transformative possibilities within family law.

This book is not a mere recitation of statutes and case law; it is a narrative woven with the threads of human experience. Each chapter unfolds like a tapestry, revealing the vibrant colors of resilience, empathy, and innovation that define the quest for family justice.

Through the adventures of Braveheart, the daring calf turned symbol of legal empathy, and the visionary insights of unrepresented children in the know, readers will embark on a journey that transcends the traditional confines of legal discourse. Here, we delve into the heart of family dynamics, exploring the triumphs and tribulations that shape our understanding of justice, innovation, and compassion.

"The Law Doctor 2" invites you to traverse the shadows of family law gaps, to confront the adversities and celebrate the triumphs, and ultimately, to envision a future where the quest for juSTEMice is intertwined with the quest for humanity. As you turn these pages, may you find inspiration, enlightenment, and a renewed sense of purpose in the pursuit of a more equitable and compassionate economy.

Introduction

Welcome to the realm where the echoes of heartache and hope intertwine, a realm we traverse in "The Law Doctor: Delivering juSTEMice to You." Here, the fabric of family law is not just a web of regulations but a canvas painted with stories of resilience, compassion, and the relentless pursuit of fairness.

Picture yourself at the crossroads of tradition and transformation, where the past whispers of struggles merge with visions of a brighter tomorrow. This odyssey isn't a mere journey through statutes and courtrooms; it's an exploration of human courage and innovation.

Follow Braveheart, the calf bearing the emblem of "Law Doctor," through daring escapades that mirror the courage needed in the quest for justice. Braveheart's visionary creations illuminate the path toward a more equitable world.

We invite you to unravel the mysteries veiled in family law's shadows, discovering tales of bravery, compassion, and the unwavering spirit of those who challenge norms. As each chapter unfolds, may you find inspiration and insight, embarking on a journey where justice isn't just a dream but a tangible reality for all. Let the odyssey begin.

Prologue

In the heart of a bustling city, where the echoes of history intertwine with the rhythm of modern life, there lies a courthouse—a sanctuary of justice, a theater of human drama. Within its walls, the fates of families are woven and unraveled, where the shadows of conflict cast their long reach.

It is here that our story begins, amidst the whispers of legal jargon and the weight of decisions that shape lives. Picture the scene: lawyers in solemn robes, judges presiding with impartial authority, and families awaiting verdicts that will alter their destinies.

Yet, beyond the formalities of legal proceedings, there exists a realm of hope and possibility—a realm where innovative solutions and compassionate understanding illuminate the shadows that often cloud family law. It is a realm where the voices of children, parents, and advocates echo with a shared vision of a better tomorrow.

As we step into this world of complexity and compassion, let us remember that every case is not just a legal puzzle but a human story—a story of love, loss, resilience, and the enduring quest for justice. Join us as we navigate the shadows of family justice, seeking to bring light to those who need it most.

Chapter 1

Braveheart

A Calf's Yearning

On a bright and sunny Saturday morning, young Aaron embarked on an exhilarating journey with his grandpa Wayne to the Central Illinois livestock sale barn. This place was no ordinary barn; it was a magical hub where farmers brought their mighty cattle to be auctioned off to eager buyers. The air was filled with anticipation as meat producers and ranchers gathered to bid on these majestic creatures.

As they arrived, Aaron watched in wonder as each cattle farmer, including his grandpa, skillfully backed up their trucks to the corrals of the barn to unload their livestock. The cattle, massive and magnificent, were corralled together and then paraded through a labyrinth of gates into the front of the auction area, where everyone could catch a glimpse of them.

The day continued to unfold, and Aaron, perched alongside his grandpa and their fellow farmers, could sense the excitement building. The barn was a bustling hive of activity – the lively chatter of farmers, the rhythmic sound of hooves, and the energy of the bustling crowd filled the air. Aaron's eyes sparkled with curiosity as each farmer proudly presented their cattle, transforming the auction into a splendid parade of gentle giants.

But on that extraordinary day, fate had something special in store. A spirited calf, brimming with daring and adventure, decided it was time to break free from the ordinary routine. Instead of following the predictable path like its fellow cattle, this fearless calf embarked on a daring escapade.

With graceful agility, the calf found its way to a sturdy wooden fence, as if it held a secret plan all along. In a breathtaking twist, it climbed the fence and boldly perched atop it, even straddling the auction booth's top like a fearless acrobat. From this elevated vantage point, it surveyed the entire barn with the regal confidence of a king overlooking his realm.

The auctioneer and his helpers were left in stunned silence, and a hush descended upon the barn. Every eye in the place was fixed in awe on this audacious calf, as it stood its ground, refusing to blend into the crowd.

In a quick stroke of ingenuity, a farmhand opened a side gate, and the calf sauntered out with an air of casual nonchalance, as if it had decided it had had enough of the limelight. It strolled down the sidewalk, drawing chuckles and smiles from all those who were lucky enough to witness this extraordinary feat.

As Aaron and his grandpa shared a hearty laugh with the crowd, young Aaron couldn't help but ponder the deep parallels between this daring calf's adventure and life itself. He began to realize that just when you think you have everything figured out, life has a way of tossing unexpected twists your way.

You might be wondering how this calf's escapade relates to families. Well, as Aaron grew older, he learned that life for kids and parents could be just as unpredictable as Braveheart's escape. Families often found themselves facing unforeseen challenges, much like that daring calf had encountered an unexpected situation.

But here's where the real life in new motion unfolded: Just as Braveheart found a way to navigate its unexpected predicament, Aaron discovered that kids and parents alike could unearth creative solutions to tackle the twists and turns in their lives. Sometimes, all it took was a dash of courage, a sprinkle of cleverness, and a heap of love to make things better for their families.

The Not Closed, Dream Open Law Quad Doctors AI

In the realm of family law, where complexities abound and justice often feels elusive, a new dawn was breaking. Dr. Amelia Harrison, a visionary legal scholar and computer scientist, was pioneering the way forward. She had a dream—a dream of a more just and compassionate family law system.

Dr. Harrison was unlike any other legal scholar. With her keen intellect, unwavering determination, and a heart that beat for justice, she embarked on a quest to revolutionize family law. She envisioned a system where families could find resolution without being torn apart by adversarial legal battles.

Her dream took the form of a revolutionary creation—a highly advanced AI known as "Law Doctor." This AI, however, was unlike any other. It possessed a deep understanding of human emotions, intricate family dynamics, and the complexities of the legal system.

Law Doctor was designed to be an empathetic and impartial mediator, guiding families through the maze of family law with compassion and wisdom. It could listen to the concerns of parents, empathize with the fears of children, and analyze legal complexities with unmatched precision.

But what set Law Doctor apart was its commitment to the principle of "Dream Open Law." Dr. Harrison believed that the law should not be a closed book accessible only to legal experts. Instead, she envisioned a world where families could easily understand their rights and responsibilities.

Law Doctor was designed to be accessible to everyone, breaking down complex legal jargon into plain language that parents and children could comprehend. It offered guidance, explained legal processes, and helped families navigate the often intimidating world of family law.

As word of Dr. Harrison's visionary creation spread, families from all walks of life began to seek the counsel of the Law Doctor. It became a trusted companion for those facing family challenges, offering not only legal insights but also a compassionate ear to listen.

The Quest for Justice 2.0 Begins

Braveheart's daring escapade and Dr. Harrison's visionary creation marked the beginning of a remarkable journey. In the pages that follow, we will delve deeper into the lives of those who dared to challenge the status quo of family law. We will follow their quest for justice, their unwavering determination to create a more compassionate system, and the extraordinary transformations that unfolded along the way.

As we venture further into this chapter, remember the spirit of Braveheart and the dream of Dr. Harrison. Together, they illuminate the path toward a more just and equitable family law system—a system that understands the yearning for justice, embraces the complexities of family life, and opens the doors of understanding to all.

Discover why every hero might just need a cape!

So, my young friend, let us embark on this journey of discovery, where each step brings us closer to unraveling the tapestry of family law and revealing the heroes who dare to challenge its norms and seek a brighter, more just future.

[To be continued...]

Chapter 2

The Art of Playing Family Law vs. Law Gaps, And Double Previously Unrepresented Family Law 2.0 Rescues

America, we find ourselves at a pivotal moment, faced with a stack of challenges that demand our attention. Among these challenges is the need for a transformative approach to family law—an approach that bridges the gaps, empowers the previously unrepresented, and champions the principles of justice and empathy.

Reflecting on the journey of Braveheart, the young law doctor who defied expectations and became a symbol of resilience, we see the potential for a future where individuals like Braveheart can evolve into teams of democratically empowered advocates. Just as Braveheart stood up against the odds, we too can stand together to navigate the complexities of family law, ensuring that justice and fairness prevail for all families, from cradle to grave.

The gates of dictators present only two stark options: confinement to the familiar farm or being sold to the slaughterhouses. However, democratized buffers offer new and improved childproof options, securities that otherwise do not exist for the democratically inclined from before the cradle until after the grave. These tracks are already in place, waiting to guide us towards a more equitable and just future.

Throughout history, pioneers and visionaries have dared to question the status quo and endeavor to redefine the boundaries of family law. Individuals like Clara Shortridge Foltz, the first woman to pass the California Bar Examination, shattered gender barriers in the legal profession, advocating for equal rights and opportunities within family law. Foltz's legacy paved the way for countless women to pursue careers in family law and champion the cause of justice.

In a different era, Rosalind Wiseman, through her work "Queen Bees and Wannabes," provided invaluable insights into the intricacies of teenage girl dynamics, shedding light on the impact of family law on adolescents. Her work inspired a generation of parents and educators to navigate the complexities of family law with a deeper understanding of its implications on young lives.

These are just a few examples of those who have stepped forward to dissect the multifaceted canvas of family law, each contributing to its ongoing evolution. In this chapter, we shall unravel the myriad threads that compose this intricate tapestry and glimpse the remarkable stories of those who have made a difference. Welcome to a journey where we delve into the very soul of family law, exploring its rich textures and seeking to illuminate the path toward a more equitable and just system.

The Canvas Unveiled

Family law is often portrayed as a profound and complex canvas, one that serves as the backdrop for the myriad stories of individuals and families. This canvas unfurls across a vast spectrum of human experiences, ranging from the elation of wedding vows to the heartache of divorce, from the boundless joys of parenthood to the arduous battles for child custody. It captures the essence of human connection and paints the intricate tapestry of human emotions.

At its essence, family law is a dynamic field of legal practice that strives to define, safeguard, and regulate the intricate web of relationships and responsibilities that bind families together. It encompasses a diverse array of legal matters, including the sacred institution of marriage, the complexities of divorce, the delicate dance of child custody, the act of adoption, the consideration of spousal support, the division of shared assets, and much more. Yet, family law transcends the boundaries of the courtroom; it weaves itself into the very fabric of our society, and in doing so, molds the destinies of generations yet to come.

For instance, consider the tale of Emma and Michael, two individuals who found themselves at the crossroads of life's journey. Their story, though unique in its details, resonates with the many narratives that converge under the umbrella of family law. Emma and Michael, deeply in love and eager to embark on a lifelong adventure together, celebrated their union with a joyous wedding ceremony. The promises they exchanged that day were more than mere words; they represented a commitment to building a life together, filled with shared dreams and aspirations.

However, as time unfurled its tapestry, Emma and Michael found themselves facing the formidable challenge of divorce. Their love had not waned, but the pathways they envisioned had diverged. In navigating the complexities of divorce proceedings, they leaned on the guidance of family law to ensure a fair and just resolution. The legal framework of family law, with its emphasis on equitability and the well-being of children, provided the scaffolding for their journey through this tumultuous chapter.

In the realm of family law, the stories of Emma and Michael are just one brushstroke on the expansive canvas of human existence. Each narrative is unique, yet interconnected by the common thread of family bonds and the intricacies of personal relationships. As we delve into the chapters that follow, we will journey through the labyrinthine corridors of family law, exploring its challenges, triumphs, and transformative power. Together, we shall uncover the secrets of Contract Deprivation Zones (CDZs) and the revolutionary solutions offered by Childproof Deal Data Analytics (CDAs). These concepts, forged through rigorous scientific inquiry, hold the potential to reshape family justice and ensure a brighter future for all. Welcome to "Law Doctor II: Navigating the Shadows of Family Justice," where the canvas of family law unfurls before us, awaiting the strokes of innovation and justice to paint a brighter tomorrow.

The Artist's Palette

Family law, like an artist's palette, is rich with colors representing the diverse array of human experiences. Each case is a unique blend of emotions, circumstances, and legal considerations. As family law practitioners, lawyers wield the brushes of advocacy, negotiation, and litigation to paint a picture that best serves their clients' interests.

But family law is not solely the domain of attorneys; it is a field that engages the collective efforts of customers, judges, mediators, counselors, social workers, and policymakers. Together, they work to create a legal landscape that balances the rights and responsibilities of individuals with the best interests of families and children.

The Human Stakes

In every family law case, there are human stakes that cannot be understated. The decisions made within the family law system can alter the course of lives, influencing the well-being of spouses, children, and extended family members. The consequences of these decisions can resonate for years, if not generations, impacting financial stability, emotional health, and the dynamics of familial relationships.

It is within the context of family law that we encounter stories of resilience and resolve, of love and heartbreak, of triumphs and tribulations. It is a domain where human vulnerability is laid bare, and where the quest for justice is intertwined with the complexities of human nature.

The Landscape of Challenges

While family law seeks to navigate the intricate terrain of human relationships, it is not without its challenges. The adversarial nature of divorce proceedings, the emotional toll of child custody battles, and the complexities of property division are but a few of the hurdles that individuals and families face within the family law system. Moreover, family law must grapple with evolving societal norms and the need to adapt to the changing dynamics of modern families.

In the second half of this chapter, we delve deeper into these challenges and explore the quest for a more just system. We will examine the shadows that loom over the landscape of family law and shed light on innovative approaches and transformative solutions that offer hope for a brighter future.

But first, let us continue our journey into the complex canvas of family law, understanding its intricacies and its impact on the lives it touches.

The Victim Endangering Shadows of Family Law

One of the most prominent shadows in family law is the adversarial nature of divorce proceedings. When spouses choose to part ways, the legal process often pits them against each other in a battle for assets, custody, and financial support. What should ideally be a process of amicable separation can escalate into a protracted and emotionally taxing ordeal.

Consider the case of Sarah and David, a couple whose love had once burned brightly but had now dimmed to a flicker. Their decision to divorce was not taken lightly, yet the adversarial nature of the legal proceedings cast a dark cloud over their attempts to part ways amicably. What should have been a time for healing and reflection turned into a battleground of accusations and counter-accusations. Legal mandates compelled them to engage in contentious debates, file motions against each other, and engage in a destructive cycle of litigation. The very system designed to facilitate their separation ended up exacerbating their pain and conflict.

Child custody battles, another shadow on the canvas, can leave lasting scars on families. The well-being of children is at the heart of these disputes, yet they can become a battleground where parents vie for control. Consider the case of Lisa and John, parents of two young children, who found themselves embroiled in a bitter custody battle after their marriage dissolved. The complexity of determining what is in the best interests of the children often leads to prolonged litigation, causing emotional distress not only to Lisa and John but also to their innocent children caught in the crossfire. The fear of missing out (FOMO) on crucial developmental milestones or a stable family environment due to prolonged legal battles compounds their pain, making it difficult for them to co-parent effectively.

Property division, too, casts its shadow over divorcing couples. Deciding how to fairly divide assets acquired during the marriage can become a contentious issue, especially when substantial wealth or complex assets are involved. Consider the case of Michael and Emily, a couple whose marriage had once been filled with dreams of a shared future. As they faced the prospect of divorce, their dreams crumbled, and they were left to grapple with the division of their accumulated assets. Legal mandates compelled Michael and Emily to engage in a prolonged and adversarial process, depleting their financial resources and leaving them emotionally drained. The fear of missing out on financial stability or a fair resolution further exacerbates their distress.

These examples underscore the harsh realities faced by individuals navigating the complexities of family law. The shadows cast by adversarial legal processes, contentious child custody battles, and acrimonious property divisions can engulf families in turmoil. It is imperative that we seek innovative solutions to dispel these shadows and usher in a new era of family justice, one where families can heal and rebuild without being subjected to further harm.

Innovative Approaches and Transformative Solutions

Amidst these challenges and shadows, there is hope. Family law is not stagnant; it is a field that is evolving to meet the needs of contemporary families. Innovative approaches and transformative solutions are emerging to address the shortcomings of the current system.

Democratized mediation and collaborative guardian divorce processes offer alternatives to traditional litigation, promoting cooperation and reducing conflict. These methods prioritize open communication and negotiation, empowering couples to reach mutually beneficial agreements.

Organic parenting plans and 360-degree shared custody arrangements prioritize the best interests of the child by fostering ongoing parental involvement and co-parenting corporations. By providing clear guidelines for new co-parenting, these arrangements aim to minimize disputes and promote stability for children who may feel doubly pressured and underrepresented.

Technological advancements are also playing a pivotal role in shaping the future of family law. Online platforms and apps are streamlining administrative tasks and improving access to legal information. Artificial intelligence is being explored to enhance case management, well-being, and streamline legal processes, ensuring a more efficient and effective system for all parties involved.

Meeting the Heroes

In the midst of this evolving landscape, we encounter remarkable heroes—children and parents who face unique challenges within their families and communities. Their stories illuminate the possibilities of creative solutions and inspire us to envision a more just family law system.

Madison and Chris: A Dynamic Duo

Madison was a bright and caring kid who loved spending time with her dad, Chris. They laughed together, shared stories, and created special memories. But one day, something happened that changed their lives. It was a problem called "Continuance Confusion," which was part of the tricky Puzzle of Prolonged Parent Separation (PPPS).

You see, sometimes parents and kids like Madison and Chris face a challenge. It feels like time keeps slipping away, and they can't spend as much time together as they want. Madison didn't like it when she had to be apart from her dad for too long. It made her sad, and she missed him dearly. She wished there was a way to fix this problem, to make sure they weren't apart for too many days.

That's when Madison decided to become a hero in her own way. She called her plan "Madison's Initiative." With the help of her dad, Chris, they wanted to reduce the time they spent apart because of "Continuance Confusion." They knew that spending time together was essential for their happiness and love as a family.

Madison's Initiative was like a magical tool that could help other families too. It was a solution known as "Madison's Time-Saver." This special solution made sure that families like Madison and Chris didn't have to be apart for too many days. It brought parents and kids closer, just like it should be.

Madison and her dad, Chris, became superheroes in their own story. They knew that even when faced with challenges like "Continuance Confusion," they could find solutions and make things better for their family.

Lily: A Heart of Gold

Lily was a sweet and cheerful girl who loved playing with her friends and spending time with her family. She had a fluffy cat named Whiskers, who was her best friend. They did everything together, from playing hide-and-seek to cuddling up for bedtime stories.

But one day, Lily's family faced a big problem. Her parents, Mr. and Mrs. Johnson, were going through something called "Continuance Confusion" (CDZ). It meant that they had to go to a special place called a courthouse because they needed help with some grown-up problems. Lily didn't quite understand all of it, but she knew that it made her parents sad, and that made her sad too.

Lily missed spending time with her mom and dad. They used to go on picnics, bake cookies, and have fun adventures together. Now, they seemed busy all the time, and Lily felt like she was missing out on all the good times they used to have.

One sunny day, as Lily sat on her porch, she saw Madison and her dad, Chris, walking by. Madison waved with a big smile, and Lily waved back. Madison and Chris were like neighborhood heroes, always helping families when they needed it most. Lily knew they were special because they used something called "Childproof Data Analytics" (CDA) to understand and solve problems.

Lily decided to talk to Madison about her family's CDZ. Madison listened with a caring heart, and she understood how much Lily missed her mom and dad. She promised to help.

With the power of CDA, Madison and Chris worked with Lily's parents to find a solution. They created a special schedule that allowed Lily to spend more time with her mom and dad, even during the grown-up problems they were facing. It was like piecing together a puzzle, and Madison and Chris made sure everyone's hearts felt lighter.

Alayna: A Shining Star

In the same friendly community, there lived another amazing kid named Alayna. Alayna was like a bright star in the night sky, always shining with happiness. Her dad, Aron, was her partner in fun and adventure.

Alayna and Aron loved spending time together, especially talking on the phone when they had to be apart. Their phone calls were like treasures, full of laughter and love. But not all families were as lucky as them when it came to phone time.

One day, while Alayna and Aron were chatting on the phone, Alayna overheard her dad talking about a big problem. It was called "Phone Time Predicament" (CDZ), and it made her feel sad. This problem meant that some families had trouble talking to their loved ones on the phone. It was like they had a magical treasure chest, but they couldn't open it to share their precious chats.

Alayna couldn't bear to think of families missing out on special phone time, just like the ones she and her dad enjoyed. So, she decided to be a hero, just like in her favorite stories.

That evening, she said to her dad, "Daddy, we have to do something to help those families who can't talk on the phone. We can be heroes!"

Aron smiled at his amazing daughter and said, "You're right, Alayna. We can use something special called 'Aron & Alayna's Call Saver' (PG-13 CDA 2). It's like a magic spell that ensures families can keep their phone chats safe."

With their newfound magic, Alayna and Aron set out to help families in need. They met parents and children who missed talking to each other because of complicated rules. It was like solving a big puzzle, and Alayna loved puzzles!

They used "Aron & Alayna's Call Saver" to make things better. It was like opening the magical treasure chest for these families. Phone time became fun and easy, just like it should be.

Alayna realized that being a hero wasn't just about having superpowers. It was about caring for others and finding solutions to their problems. She and her dad were doing just that, making the world a brighter place, one phone call at a time.

Ellie and Evan: Sibling Heroes

In a cozy corner of the same neighborhood, there lived two extraordinary kids, Ellie and Evan. These siblings were as close as two peas in a pod, and they shared a special bond with their dad, Denis.

Ellie and Evan's dad was a wise and caring man who loved them more than anything in the world. But even the happiest families face challenges sometimes, and theirs was no different. They encountered a problem known as the "Toxic Talk Trap" (PG-13 CDZ 4), which was like a dark cloud hanging over their home.

The "Toxic Talk Trap" was a tricky problem where parents argued a lot, and it felt like there was no end to their disagreements. Ellie and Evan didn't like it one bit when their home felt filled with tension and worry. They wanted their family to be a place of love and laughter, just like it used to be.

One evening, after a particularly tough argument between their parents, Ellie and Evan decided it was time to be heroes. They didn't have capes or superpowers, but they had something even better—love and determination.

Evan, being the older sibling, had a brilliant idea called "Evan's Initiative." It was like a magic spell that helped grown-ups talk about their disagreements while keeping the children's feelings safe and sound. And Ellie had her own secret weapon too, something she had learned from her dad's special book about Elizabithia.

Elizabithia was a magical kingdom where seven varieties of trees known as Harmonic Timbers grew. Each of these trees emitted unique frequencies, creating seven special notes that formed the basis of all the music in Elizabithia. But there was a secret hidden in an ancient scroll—a secret that Evan's dad, Denis, had shared with Ellie.

This secret spoke of the possibility of other Harmonic Timbers, ones that produced frequencies unknown to the citizens of Elizabithia. Evan's dad, Denis, had enlisted the help of his sister, Ellithia, and his good friend, Dionysius, to search for these lost music trees. Together, they were known as the Trio.

Following the information from the ancient scroll, the Trio embarked on a journey to a land known as Cymatigonia in search of the lost music trees. Their quest would shake up the music world, but it wouldn't be easy. They faced resistance from Elizabithian authorities who opposed their mission.

But just like Evan's Initiative helped their parents talk more kindly, the Trio had the courage to pursue the truth about music. They knew that there was far more to music than anyone had realized, and they were determined to prove it.

As they journeyed through Cymatigonia, they contended with long-held musical prejudices and sought to show that the world of music was broader and more beautiful than anyone had ever imagined. The Trio knew the truth, and their adventure would change the history of music forever.

As we meet these extraordinary heroes and their families, we must ask ourselves: How can their stories inspire us to envision a more just family law system, one that minimizes conflict, prioritizes the well-being of children, and adapts to the ever-changing landscape of family dynamics?

Their journeys hold valuable lessons and offer glimpses of innovative solutions that can transform the complex canvas of family law into a more equitable and compassionate landscape.

The Quest for a More Just STEM-Inclusive System

In the world of family law, as with any canvas, shadows are cast amidst the vibrant colors of human experience. These shadows compel us to seek a more just and compassionate system. As we delve deeper into the complexities of family law, we must confront the challenges that have plagued this field and endeavor to find innovative solutions that can lead us toward a brighter future.

The Adult Shadows of Family Law

One of the most prominent shadows in family law is the adversarial nature of divorce proceedings. When spouses choose to part ways, the legal process often pits them against each other in a battle for assets, custody, and financial support. What should ideally be a process of amicable separation can escalate into a protracted and emotionally taxing ordeal.

Child custody battles, another shadow on the canvas, can leave lasting scars on families. The well-being of children is at the heart of these disputes, yet they can become a battleground where parents vie for control. The complexity of determining what is in the best interests of the child can lead to prolonged litigation, causing emotional distress for all involved.

Property division, too, casts its shadow over divorcing couples. Deciding how to fairly divide assets acquired during the marriage can become a contentious issue, especially when substantial wealth or complex assets are involved. The process often requires extensive legal battles, leaving both parties financially depleted.

Evolving Family Dynamics

The canvas of family law is further complicated by evolving societal norms and family structures. Traditional notions of marriage and parenthood no longer define the boundaries of family life. Blended families, same-sex couples, and cohabiting partners have challenged the conventional definitions that once guided family law.

As our understanding of family dynamics evolves, so too must the legal framework that governs them. Family law must adapt to accommodate the diverse array of family structures, ensuring that all families receive equitable treatment under the law.

Innovative Approaches and Transformative Solutions

Amidst these challenges and shadows, there is hope. Family law is not stagnant; it is a field that is co-evolving to meet the needs of contemporary families. Innovative approaches and transformative solutions are emerging to Constitutionally re-dress the shortcomings of the current system.

Democratized mediation and collaborative divorce processes offer alternatives to traditional litigation, promoting cooperation and reducing conflict. These methods prioritize open communication and negotiation, empowering couples to reach mutually beneficial agreements.

Systemic plans and parenting plans with shared collaborative arrangements seek to prioritize the true best interests of the child by fostering ongoing co-parental involvement, measurements, and cooperation. By providing clear guidelines and gauges to follow for healthful co-parenting, these #newDeal terms aim to minimize disputes before they become litigious and promote organic stability for children, homes, and families.

Technological advancements are also playing a role in shaping the future of family law. Although at times risky, online platforms and apps are streamlining organic and administrative tasks to improve access and adaptations to legal information. Artificial intelligence is being explored to enhance case management and streamline legal processes.

Don't miss out on revolutionizing family law – your family could be next!

As we navigate the complex canvas of family law and its shadows, one question lingers: How can we create a Family Law 2.0 network that truly serves the needs of individuals, conventional family law, and families, one that co-minimizes conflict, prioritizes the well-being of children, and adapts to the ever-changing landscape of family dynamics?

In the chapters that follow, we will embark on a journey to explore the possibilities. We will meet individuals and organizations that are dedicated to reimagining family law, seeking to create a system that is more just, compassionate, and responsive to the diverse needs of families. Our quest for a brighter future in Family wellbeing begins now.

Chapter 3

In Search of JuSTEMice - Defining juSTEMice

In our journey to unravel the complexities of family law and seek a more just and equitable system, we must first delve into the concept of JuSTEMice. This term represents a profound shift in the way we approach family law—a shift that is rooted in science, technology, engineering, and mathematics (STEM) principles and guided by the philosophy of performance parenting.

The Traditional Landscape

Traditionally, family law has been viewed through a binary lens—dividing families into winners and losers, plaintiffs and defendants, and custodial and non-custodial parents. This binary perspective has often resulted in a zero-sum game, where one party's gain is perceived as the other's loss. It has fostered an adversarial environment that pits family members against each other, with the legal system acting as the referee.

The JuSTEMice Paradigm Diversity

JuSTEMice represents a departure from this binary paradigm. It envisions family law as a dynamic system, where science and technology play a pivotal role in guiding decisions and resolutions. Here's how JuSTEMice redefines family law:

1. Organic Data vs. Systemic Words

In the traditional family law system, decisions are often based on words—testimonies, arguments, and legal documents. These words form the basis of judgments, leaving room for misinterpretation and subjective biases. JuSTEMice, on the other hand, relies on organic data.

Organic data encompasses a wide range of objective, quantifiable information—financial records, communication histories, parenting schedules, and more. It paints a comprehensive picture of a family's dynamics, leaving little room for ambiguity or manipulation. By embracing data-driven decision-making, JuSTEMice ensures that resolutions are based on concrete evidence rather than persuasive rhetoric.

2. Records, Scores, Metrics, and Incentives

JuSTEMice introduces a systematic approach to family law, where every aspect is quantified, measured, and evaluated. This involves creating records of family interactions, assigning scores to parenting performance, tracking metrics related to child well-being, and implementing incentive systems that reward positive behaviors.

For example, rather than relying solely on subjective assessments of a parent's suitability, JuSTEMice utilizes a scoring system that considers factors such as stability, emotional support, and financial responsibility. These scores provide an objective basis for decisions, ensuring that children's best interests remain at the forefront.

3. 360-Degree Performance Parenting Facilitator

Performance parenting, a key component of JuSTEMice, adopts a holistic approach to parenting. Instead of focusing on custody arrangements, it considers the well-being of children from all angles. This 360-degree perspective examines physical, emotional, educational, and social aspects of parenting.

JuSTEMice recognizes that children thrive when they have access to both parents and a supportive network. It encourages parents to collaborate in a shared parenting model, where responsibilities are distributed equitably, and both parents are actively involved in their children's lives.

The Path to Transformation

The transformation from traditional family law to JuSTEMice is not a simple one—it requires a fundamental shift in mindset, processes, and technology. Here's the path that leads us to this transformative paradigm:

1. Data Collection and Analysis

At the core of JuSTEMice is data. Families and individuals would engage in the systematic collection of data related to their interactions, financial contributions, and parenting responsibilities. Technology would play a crucial role in automating data collection, ensuring accuracy, and maintaining privacy.

Data analysis tools would then process this information, generating objective insights that guide decision-making. Algorithms would consider a multitude of factors, from a child's emotional well-being to a parent's financial contributions, to produce a comprehensive view of a family's dynamics.

2. Democratized Mediation and Collaboration

JuSTEMice promotes democratized mediation and collaboration as primary methods for resolving family disputes. Rather than resorting to adversarial court proceedings, families would engage in facilitated discussions aimed at finding mutually beneficial solutions.

Co-mediators, guided by the data and victimproof, un-divorce metrics provided by JuSTEMice, would help parents reach agreements that prioritize their children's best interests. Incentives tied to positive outcomes would motivate parents to cooperate and maintain a healthy co-parenting relationship.

3. Continuous Monitoring and Adjustments

JuSTEMice is a dynamic system that adapts to changing circumstances. Continuous monitoring of family dynamics, well-being metrics, and parenting performance ensures that interventions can be made when needed.

For instance, if a child's academic performance declines or emotional well-being is affected, JuSTEMice would trigger a review of parenting arrangements. This proactive approach prevents issues from escalating and promotes a responsive system that prioritizes children's welfare.

4. Education and Support

To succeed in the world of JuSTEMice, families would receive education and support. Parents would be empowered with knowledge about effective co-parenting strategies, financial responsibility, and emotional support.

Technology-driven platforms and resources would be readily accessible, offering guidance on how to navigate the JuSTEMice system effectively. Support networks would also be in place to assist families in times of difficulty.

5. Ethical Considerations

As we navigate this transformative journey, ethical considerations remain paramount. Privacy, transparency, and fairness must be upheld at all times. The responsible use of data and technology ensures that

JuSTEMice remains a just and equitable system. However, national engineering (practitioners of STEM laws) ethics codes help.

The Faces of JuSTEMice

To truly understand the transformative power of JuSTEMice, we need to hear the stories of families who have embarked on this journey. These are stories of resilience, cooperation, and unwavering commitment to the well-being of their children.

The Andersons: A Tale of Financial Clarity

Meet the Andersons—Sarah and David. They were once caught in the web of financial disputes that often plague divorcing couples. In the traditional family law system, their arguments about child support and alimony seemed endless, with neither party satisfied.

Enter JuSTEMice, and their story takes a different turn. The Andersons realized that by embracing financial transparency, they could create a better future for their children. They began using data-driven tools that tracked their financial contributions and ensured that their children's needs were met.

With a clear understanding of their financial responsibilities and the support of JuSTEMice mediators, Sarah and David shifted their focus from disputes to collaboration. They found a fair and sustainable financial arrangement that allowed them to provide for their children without constant tension.

The Carters: Navigating Emotional Terrain

Emotions often run high in divorce and separation, and the emotional well-being of children can be deeply affected. The Carters, Jessica and Michael, understood the importance of emotional support for their two children, Emily and Ethan.

In the traditional family law system, their disputes over parenting time and emotional support were draining. But when they adopted

JuSTEMice principles, they began to view their children's emotional health as a quantifiable metric.

JuSTEMice encouraged the Carters to prioritize open communication and emotional support. They used data to track the emotional well-being of their children and adjusted their parenting strategies accordingly. Jessica and Michael learned to identify signs of emotional distress and collaborated on providing the necessary support.

Their story is one of transformation, where the emotional needs of Emily and Ethan became the central focus. The Carters embraced the idea that emotional health was as vital as physical health, and JuSTEMice provided the tools to make it a reality.

The Parkers: Co-Parenting Reimagined

Co-parenting can be a challenging journey, especially when geographical distances create barriers. The Parkers, Amanda and Mark, faced this obstacle when Mark's job required him to relocate to a different state.

In the traditional family law system, their co-parenting arrangement seemed impractical. But JuSTEMice offered a solution. By using technology to facilitate remote co-parenting, the Parkers found a way to maintain a strong bond with their children, Lily and Ethan.

Virtual visitation schedules, video calls, and shared digital calendars became integral parts of their co-parenting routine. JuSTEMice tools ensured that both Amanda and Mark were actively involved in their children's lives, despite the physical distance.

The Parkers' story showcases the adaptability and innovation that JuSTEMice brings to co-parenting. It reimagines the possibilities of maintaining strong parent-child relationships, no matter the geographical challenges.

Dive into the justice of the future – be ahead of the curve!

As we explore the stories of families who have embraced JuSTEMice, we witness a profound transformation in the world of family law. The shift from adversarial battles to data-driven cooperation is not only possible but also essential for the well-being of children and the families they belong to.

In the chapters that follow, we will delve deeper into the practical applications of JuSTEMice, examining the tools and technologies that make it a reality. We will also explore the ethical considerations and potential challenges on this transformative journey.

Our quest for JuSTEMice is a journey toward a future where families are empowered, conflicts are resolved with compassion, and the best interests of children are upheld. Join us as we continue our quest for JuSTEMice—a future where science, technology, engineering, and mathematics converge with compassionate co-parenting, creating a more just and equitable world for families.

Chapter 4

Braveheart's Quest for Double Border Gate Relief

Join the quest that's changing borders and lives!

Braveheart and his companions ventured back into the animal kingdom, eager to share the knowledge they had gained in the world of JuSTEMice. They knew that their journey would be met with challenges, but they were fueled by a determination to make a difference.

As they traveled through the animal kingdom, they encountered children who faced their own set of challenges, much like Madison, Lily, Alayna, Ellie, and Evan. These children had found themselves in the shadows of Contract Deprivation Zones (CDZs) and were often caught in the web of family disputes. But these young minds were not content with being passive victims of circumstance; they were ready to become agents of change.

One such child was Alex, a teenager with a knack for technology. He had seen his parents struggle with the complexities of family law and recognized the need for a better way. Inspired by Braveheart's journey, Alex decided to become a young JuSTEMice pioneer.

With the help of his tech-savvy friends, Alex developed a Family Law 2.0 CDA (Child Data Analytics) platform. This innovative tool allowed children and parents to record their thoughts, feelings, and concerns in a safe and confidential space. It gave them a voice in the decision-making process, ensuring that their perspectives were not lost in the legal maze. Alex's initiative sparked a sense of urgency in other young minds, fearing they might miss out on shaping the future of family law.

Another remarkable child was Mia, a budding scientist with a passion for understanding human behavior. She had witnessed the toll that prolonged parent separation took on her friends and their families. Determined to make a difference, Mia turned to the principles of JuSTEMice for guidance.

Mia, with the support of her family and friends, designed a "Parenting Harmony Index." This tool used data to assess the emotional well-being of children during family disputes. It provided valuable insights to parents, helping them make decisions that prioritized their children's happiness and stability. Mia's fear of missing out on creating impactful change fueled her determination to innovate.

Child Safety File Agents

As more children like Alex and Mia joined the ranks of young JuSTEMice pioneers, a new term emerged to describe their role: Child Safety File Agents. These young agents were not just passive participants in their family's legal proceedings; they were active contributors, using technology and data to ensure their voices were heard.

Child Safety File Agents, armed with their Family Law 2.0 CDAs, helped families navigate the complex world of family law. They facilitated open and honest communication between parents and children, fostering an environment of collaboration and understanding. Their sense of urgency stemmed from a fear of missing out on the opportunity to shape a fairer system for all.

These young double gate buffers were no longer confined to the shadowy pinch-points of CDZs; they had become beacons of hope and innovation. They were rewriting the narrative of family disputes, turning them into opportunities for growth and transformation.

Discovering the Power and Inclusivism of JuSTEMice

As Braveheart and his newfound friends delved deeper into the world of JuSTEMice, they were met with a breathtaking landscape of innovation and collaboration. The air was filled with the hum of technology, and colorful data streams crisscrossed the sky.

Their first stop was the "Data Oasis," a place where information flowed freely, much like a crystal-clear stream. Here, they learned how data could be harnessed to understand the needs and desires of families. It was a revelation that families could use data to make informed decisions, just as farmers used data to tend to their crops. The potential of this discovery instilled a sense of urgency in Braveheart and his companions, fearing they might miss out on utilizing such powerful tools for positive change.

Next, they visited the "Innovation Grove," a forest teeming with cutting-edge technology. From virtual reality simulations that allowed parents to better understand their children's perspectives to advanced communication tools that bridged the gap between separated parents, the Innovation Grove showcased the endless possibilities of JuSTEMice. The thought of missing out on these advancements fueled their determination to learn and adapt.

But Braveheart and his companions soon realized that JuSTEMice wasn't just about technology. It was about the human element—the empathy and compassion that underpinned every innovation. They met with "Emotion Engineers" who specialized in understanding the emotional needs of children and parents. These experts used their knowledge to create solutions that went beyond numbers and algorithms, reminding Braveheart and his friends of the importance of not missing out on the human touch amidst technological progress.

As Braveheart and his friends continued their journey, they encountered the "Bridge of Collaboration," a magnificent structure that spanned a vast chasm. It symbolized the coming together of families, professionals, and technology to create a brighter future. The sight of this bridge reinforced their fear of missing out on the potential harmony that collaboration could bring.

Braveheart's Mission

Braveheart, now fully immersed in the world of JuSTEMice, understood the importance of his mission. He had become a symbol of courage and innovation, a bridge between the animal kingdom and the world of family law. The fear of missing out on opportunities to bring positive change drove him forward.

With the guidance of Olliver and the support of his newfound friends, Braveheart embarked on a quest to bring the principles of JuSTEMice back to the animal kingdom. He believed that just as families could benefit from data-driven collaboration, so too could the animal kingdom find solutions to its challenges. His fear of missing out on creating a better world for all fueled his determination.

But Braveheart's journey was far from over. He knew that there were challenges and obstacles ahead. The animal kingdom was a diverse and complex place, and not everyone would embrace change as readily as he had. Yet, the fear of missing out on the opportunity to make a difference pushed him to persevere.

A Legacy of Championing The New

Braveheart, the calf who had defied convention and embarked on a quest for JuSTEMice, looked upon the children he had inspired with pride. Their courage and creativity had transformed their own CDZs into CDAs, filling the world with the promise of a more just and compassionate family law system. The fear of missing out on creating a lasting impact motivated them to continue championing innovation.

As Braveheart's journey continued, he knew that his legacy would live on through the countless children who had found their own brave hearts. Together, they were shaping a future where families could navigate challenges with empathy, understanding, and the power of JuSTEMice. The fear of missing out on contributing to this brighter future fueled their passion and dedication.

And so, the adventure of Braveheart and the young JuSTEMice pioneers continued, a testament to the boundless potential of young minds and the enduring spirit of innovation and justice.

Chapter 5

The Double Shadow, Double Valley's Mania

Navigating the Dark Alleys of Family Law

In the world of family law, there exists a place known as the Shadow Valley's Mania. It's a complex and often perplexing landscape where the lives of families intersect with the legal system. The term "Shadow Valley's Mania" encapsulates the challenges, confusion, and conflicts that can arise when families find themselves entangled in a web of legal proceedings. This ominous valley serves as a cautionary tale, reminding us of the dangers of not taking proactive measures to address the shortcomings of the existing system.

The Labyrinth of Litigation

Imagine a labyrinth, its intricate paths representing the convoluted legal processes that families must navigate when facing divorce, custody battles, and other family disputes. Each twist and turn in this maze leads to uncertainty, emotional turmoil, and financial strain.

Families are at risk of being lost in this labyrinth, succumbing to the pressures and complexities without a clear path forward.

For many, the journey through the Shadow Valley's Mania can feel like an endless ordeal. The legal system, despite its intentions, often exacerbates the very problems it seeks to resolve. Families are subjected to lengthy court battles, relentless paperwork, and mounting legal fees, all while their lives remain in a state of flux. The fear of missing out on simpler, more efficient processes haunts those who are engulfed in this legal maze.

The Emotional Toll

The toll of Shadow Valley's Mania is not limited to paperwork and court appearances; it extends to the emotional well-being of children and parents. The constant stress, conflict, and uncertainty take a profound toll on the mental and emotional health of those involved. Families risk missing out on precious moments of stability and harmony, replaced instead by anxiety and discord.

Children caught in the crossfire of family disputes often experience anxiety, depression, and feelings of abandonment. Parents, too, grapple with overwhelming stress as they attempt to secure the best interests of their children while navigating the legal labyrinth. The fear of missing out on a peaceful and nurturing environment for their children weighs heavily on their minds.

The Urgent Need for Change

The Shadow Valley's Mania is a stark reminder that the current family law system is in dire need of reform. It's a system that often perpetuates conflict rather than resolving it. The traditional adversarial approach, where lawyers on opposing sides battle it out in court, often leaves families fractured and drained. There is a palpable fear of missing out on opportunities for amicable resolutions and constructive outcomes.

A Call for Compassionate Justice

The urgent need for change in family law has given rise to a growing movement—one that seeks to replace the adversarial model with a more compassionate and constructive approach. This movement recognizes that families deserve a legal system that promotes healing, cooperation, and the well-being of all members. It's a call to action driven by the fear of missing out on transformative solutions that prioritize human dignity and harmony.

The JuSTEMice Solution

Enter JuSTEMice, a paradigm shift that introduces science, technology, engineering, and math (STEM) principles into family law. JuSTEMice recognizes that family disputes are complex and multifaceted, requiring innovative and data-driven solutions. It represents a beacon of hope for those who fear missing out on fair and efficient resolutions.

JuSTEMice empowers families with tools like Child Data Analytics (CDAs) to make informed decisions based on data rather than emotion. It emphasizes the importance of early intervention and preventive measures to address issues before they escalate into full-blown disputes. The promise of JuSTEMice is a future where families no longer fear missing out on justice and well-being.

In the grand tapestry of familial struggles and legal complexities, Braveheart, Madison, Alayna, and their companions stand as gatekeepers and guardians of a new era. Their collective journey through the valleys of shadows, akin to traversing the valleys of the shadow of death, symbolizes not only their personal growth but also the diversification of justice. Without their bravery and commitment to forging buffer options and border gateways, their lives might have been led into the abyss of missed opportunities and unfulfilled promises.

Braveheart's courage to challenge convention and embrace JuSTEMice principles serves as a beacon of hope for others trapped in the labyrinth of traditional family law. Madison's resilience and Alayna's determination to navigate emotional terrain represent the multifaceted aspects of justice that encompass emotional well-being, not just legal frameworks. Together, they embody the concept of "double justice" — not just seeking legal resolutions but also ensuring emotional and psychological harmony within families.

Had they not taken the path less traveled, their lives might have been confined to the repetitive cycles of conflict and despair seen in the Shadow Valleys Mania. The fear of missing out on a brighter future, where collaboration and empathy reign, would have loomed large over their existence. However, by embracing their roles as gatekeepers of change, they have opened doors to a world where families can thrive amidst challenges, where justice is not a singular path but a mosaic of diverse perspectives and inclusive solutions.

Stories of Transformation

To truly grasp the transformative capabilities of JuSTEMice, we must delve into the narratives of families who, like Braveheart, navigated the Shadow Valleys Mania and emerged victorious. Each story echoes the fear of missing out on a harmonious resolution, a fear conquered through innovation and collaboration.

The Smith Family

Amidst the turbulent currents of a custody dispute, the Smith family faced the looming specter of protracted legal battles and emotional strain. However, guided by a JuSTEMice-trained mediator, they charted a different course. Through meticulous data analysis, they crafted a parenting plan that prioritized their children's well-being. Their journey reflects the fear of missing out on a cooperative co-parenting model, a fear dissolved by the power of data-driven solutions.

The Garcia Family

For the Garcia family, the threat of financial and emotional depletion loomed large during their divorce settlement negotiations. Yet, inspired by JuSTEMice principles, they embraced collaborative law practices. Transparent negotiations became the cornerstone of their resolution, sparing their children the anguish of prolonged legal turmoil. Their story embodies the fear of missing out on a fair and mutually beneficial agreement, a fear dispelled by the ethos of cooperation and fairness.

Uncover the secrets of the Double Valley before they disappear!

The Shadow Valleys Mania, with its labyrinthine legal complexities and emotional toll, casts a daunting shadow over many families. But amidst this darkness, JuSTEMice emerges as a beacon of hope and innovation. In the upcoming chapters, we will unravel the transformative potential of science, technology, engineering, and math in reshaping the landscape of family law. The tales of the Smith and Garcia families are but a glimpse into a future brimming with promise—a future where families need not fear missing out on justice, collaboration, and compassion.

As we turn the page to embark on this enlightening journey through the realms of JuSTEMice, let us linger in anticipation, for the stories yet untold promise a tapestry of resilience, ingenuity, and the triumph of justice over adversity. Brace yourselves for a narrative that transcends legalities and delves into the very essence of human connection and progress. The stage is set, the characters poised for transformation—let us venture forth and witness the unfolding saga of JuSTEMice's profound impact on the fabric of family law.

Chapter 6

Concealing Vices, Devices & Divorces: CDZs Unveiled

Contract Deprivation Zones (CDZs) Explained

The world of family law resembles a complex tapestry, intricately woven from the threads of human relationships, legal agreements, and societal norms. Yet within this rich fabric, there exist fault lines—areas where the threads have frayed, leaving gaping holes that can have profound consequences. These voids are what we refer to as Contract Deprivation Zones, or CDZs.

Unveiling the CDZs

To grasp the concept of CDZs, envision a safety net akin to the one used by trapeze artists in a circus. This net is designed to catch and shield anyone who might fall during their daring performances.

However, what if this safety net had gaping holes, spaces where individuals could slip through and plummet to the ground below? These openings symbolize CDZs within family law agreements.

CDZs Defined

CDZs emerge when family law agreements lack crucial elements, manifesting in various forms:

- Financial Transparency: Instances where financial aspects like income disclosure or asset division are obscure or concealed.
- Emotional Support: Scenarios where the emotional needs of children or parents are disregarded or inadequately addressed.
- Effective Co-Parenting: Situations where agreements fail to foster effective communication and collaboration between divorced or separated parents.

The Holes in the Family Law Net

CDZs are akin to the perforations in a safety net, posing significant risks to those embroiled in family law disputes. These zones often serve as breeding grounds for conflicts, misunderstandings, and protracted legal battles.

Financial CDZs

Consider the plight of the Johnson family. Mr. Johnson, a thriving entrepreneur, and Mrs. Johnson, a devoted homemaker, had built a life together. However, upon their decision to part ways, their divorce proceedings unveiled a substantial CDZ—financial transparency.

Mr. Johnson had shrewdly concealed certain assets, leaving Mrs. Johnson oblivious to their existence. This lack of financial transparency birthed bitter disputes, prolonged courtroom sessions, and exorbitant legal fees. Ultimately, it was the children who bore the brunt, caught amidst their parents' financial strife.

Emotional CDZs

Turning our gaze to the Garcia family once more, we encounter a different CDZ—emotional support. As their divorce saga unfolded, their children, Sofia and Carlos, found themselves engulfed in emotional turmoil.

The court's fixation on asset allocation and custody arrangements often eclipsed the emotional needs of these tender hearts. The absence of emotional support mechanisms and counseling for the children left lasting wounds that would require years to heal.

Co-Parenting CDZs

Co-parenting post-divorce is a nuanced dance that requires delicate balance. Sadly, many family law agreements treat effective co-parenting as an afterthought, leading to co-parenting CDZs where communication falters, and children unwittingly become pawns in their parents' conflicts.

Braveheart's CDZ Before Buff: Financial Vulnerability

Before Braveheart embraced the principles of JuSTEMice, his family found themselves in a CDZ marked by financial vulnerability. Amidst the tumult of divorce proceedings, crucial financial information was obscured, leading to misunderstandings and contentious debates over asset division and support.

Madison's CDZ: Emotional Neglect

Madison, a young girl navigating her parents' divorce, experienced an emotional CDZ characterized by neglect. The focus on legalities overshadowed her need for emotional support and stability, leaving her feeling unheard and emotionally adrift.

Alayna's CDZ: Communication Breakdown

Alayna's family faced a CDZ rooted in communication breakdowns. Without effective co-parenting strategies or clear communication channels, misunderstandings escalated, fueling conflicts and straining familial relationships.

Evan's CDZ: Lack of Stability

Evan, amidst his parents' separation, grappled with a CDZ stemming from a lack of stability. Unclear custody arrangements and inconsistent parenting approaches left him feeling unsettled and insecure about his future.

Ellie's CDZ: Alienation and Isolation

Ellie's CDZ was marked by feelings of alienation and isolation. The adversarial nature of the legal process and parental conflicts led to her feeling disconnected from both parents, impacting her sense of belonging and emotional well-being.

Universal CDZ: Information Asymmetry

A commonly encountered CDZ, regardless of specific circumstances, is information asymmetry. When one party possesses more knowledge or access to resources than the other, it creates an imbalance of power and can lead to unfair outcomes in legal negotiations and decision-making.

Example of Universal CDZ: Home Buying

Consider the scenario of purchasing a home. If one party, say the seller or the real estate agent, withholds crucial information about the property's condition or market value from the buyer, it creates a CDZ of information asymmetry. The buyer, lacking access to vital details, may make decisions that are not in their best interest, leading to potential disputes and regrets down the line.

Reveal what's hidden—be the first to know the unseen!

The prevalence of Contract Deprivation Zones (CDZs) within family law agreements is not incidental but rather indicative of a system that, in its current configuration, often neglects the fundamental welfare of families. As illuminated through the intricate narratives of the Johnson and Garcia families, these voids can yield profound and enduring repercussions, particularly impacting the children enmeshed within such legal complexities.

In the forthcoming chapters, we shall embark on an odyssey delving deeper into the revolutionary efficacy of Child Data Analytics (CDAs) in discerning and mitigating these insidious CDZs. We shall navigate beyond the veils of divorce's shadows, ushering families toward a future where their necessities and welfare are paramount in the realm of family law.

Chapter 7

Unveiling CDZs: The Childproof Deal Data Analytics (CDAs)

A Tale of Transition: The Radiance of Childproof Data

As we embark on this illuminative journey into the realm of CDAs, let us pause to appreciate the transformative radiance they bestow upon the intricate tapestry of family law. Like luminescent beacons cutting through the murky waters of legal ambiguity, CDAs offer a beacon of clarity and precision. Imagine a world where every familial nuance, every emotional resonance, and every strategic necessity is encapsulated within the secure vaults of data-driven wisdom. Such is the promise of CDAs—the guardians of equitable resolutions and the antidote to the insidious encroachment of CDZs. Prepare to witness the unfolding saga of Childproof Deal Data Analytics, where data isn't just numbers but a narrative of familial integrity and justice.

A Beacon of Light

Envision a towering lighthouse amidst stormy seas, its resolute beam piercing the darkness, guiding vessels safely to harbor. In the tumultuous realm of family law disputes, Child Data Analytics (CDAs) emerge as modern-day lighthouses, steering families through the tempestuous waters of divorce and custody battles.

But what exactly are CDAs, and how do they work their transformative magic?

CDAs as Quantum Layer Taylors (QLTs)

The term 'Quantum Layer Taylors' may sound akin to science fiction lore, yet within family law's domain, it encapsulates a profound reality. CDAs function on a meticulously detailed level, akin to how quantum physics delves into the nuances of the subatomic realm. These Quantum Layer Taylors tailor or customize solutions, addressing the specific needs and nuances of each family, fortifying vulnerable areas that traditional family law often neglects.

Monitoring CDZs to De-Risk Them with Fitting CDAs

Think of CDAs as a vigilant shield perpetually scanning family law agreements, seeking out vulnerabilities. When a potential Contract Deprivation Zone (CDZ) is identified, CDAs spring into action, proactively mitigating risks. It's akin to using intelligent technology to anticipate weaknesses in a safety net and reinforce it beforehand.

Let's revisit the Garcia family, beset by emotional CDZs during their divorce proceedings. With the integration of CDAs, their narrative could have evolved drastically.

Step 1: Identifying the Emotional CDZ

CDAs meticulously assess the family's emotional landscape, taking into consideration the sentiments and requirements of each member.

In the case of Sofia and Carlos, CDAs swiftly pinpoint the emotional CDZ resulting from inadequate support and counseling.

Step 2: Tailoring a Solution

With the emotional CDZ identified, CDAs propose tailored remedies. This may encompass counseling sessions for both children and parents, aiding them in navigating the emotional complexities of divorce.

Step 3: Continuous Monitoring

CDAs go beyond offering recommendations; they ensure the implementation of agreed-upon measures. Regular evaluations and check-ins facilitate progress tracking and necessary adjustments. In the Garcia family's scenario, CDAs would monitor Sofia and Carlos's emotional well-being, guaranteeing they receive the requisite support.

Master the art of childproof deals and data – stay secure!

CDAs form the foundation of a '1st Chances 1st, Licensed Safer Private Network (SPN), providing families with a robust support system. This network, steered by licensed professionals utilizing organic data-driven insights, navigates families through the labyrinthine intricacies of divorce, custody, and associated disputes.

In the forthcoming chapters, we shall delve into real-world anecdotes of families benefiting from CDAs and the SPN's guidance. These narratives will illuminate how this innovative family law approach fosters fairness and justice for all parties involved.

Chapter 8

The Awakening: From Vacated to Unvacated

Have you ever experienced a sense of vacancy in crucial areas of your life? Situations where your aspirations seemed out of reach due to limited resources and support? In this chapter, we delve into a transformative journey that explores the profound impact of confirmation and connection on young minds. We contrast the stark contrast between the sense of vacancy experienced by young kids like Braveheart, deprived of essential elements, and the adaptive nature of eye-witnessing true confirmations through Real QuadCash AI Coins and parental presence, even if remotely.

The Vacancy of Absence

Young minds like Braveheart often face a deep sense of vacancy, lacking physiological, media, psychological, medical, educational, and spiritual elements crucial for holistic growth and development. This absence can lead to a disconnect from essential aspects necessary for their well-being.

RMX: Navigating the Turbulent Seas of Fate

Imagine RMX, the Rescue Market X, as a lifeboat navigating the tumultuous seas of fate. It carries within it four Real QuadCash AI Coins, representing authentic connections and validations, and two steadfast parents symbolizing unwavering guidance and support. Together, they form a shield against the perils that threaten to devour destinies, ensuring security and growth from cradle to grave.

Actionable Insights and Expertise

Drawing from expertise in family law, organizational leadership, and psychology, actionable insights emerge. Encouraging families to embrace the concept of RMX means prioritizing genuine connections, leveraging technology for meaningful interactions, and fostering resilient mindsets in children navigating today's digital age.

Untraporithm Buffers Matter

Real QuadCash AI Coins emerged as a response to the complex trials and challenges faced by individuals and families in today's digital landscape. They are designed to verify and validate online interactions, providing a secure platform for communication and transactions.

Transforming Challenges into Opportunities

In a world where cyberbullying, media biases, AI algorithms, and legal complexities pose significant threats, Real QuadCash AI Coins offer a lifeline. They act as a buffer against predatory practices, ensuring that individuals and families can navigate digital spaces without fear of exploitation or harm.

Empowering Change Through Action

Through training programs and webinars, individuals can learn to navigate digital spaces safely using Real QuadCash AI Coins. These coins break free from negative digital cycles, offering a pathway to resilience, empowerment, and security.

Visualizing a Safer Digital Future

As we envision a future where double borders and quadruple representation become the norm, Real QuadCash AI Coins stand as a beacon of hope. They transform open cliff edges into fortified safety guards, ensuring that dreams and aspirations are not hindered by digital threats or biases.

Be a part of the revival - A journey of validation and resilience

"The Awakening: From Vacated to Validated" sheds light on the transformative power of confirmation and connection in shaping the lives of young minds. Through Real QuadCash AI Coins and parental presence, even in remote settings, we witness a paradigm shift towards holistic well-being and adaptive resilience in today's digital age.

Chapter 9

Illuminating JuSTEMice: The Quad Quantum Organic Solutions

Adaptations of JuSTEMice and CDAs

In our exploration of family law's transformation, we've encountered two powerful allies: JuSTEMice and Childproof Deal Data Analytics (CDAs). These seemingly distinct forces are about to converge, unleashing a wave of innovation that will reshape the landscape of family law.

The Intersection of JuSTEMice and CDAs

JuSTEMice, rooted in the principles of science, technology, engineering, and mathematics (STEM), transcends traditional boundaries. It embraces a 360-degree approach to parenting, acknowledging the multifaceted nature of familial relationships.

On the other hand, CDAs act as vigilant guardians of family law, identifying and mitigating Contract Deprivation Zones (CDZs) to protect the interests of children and parents alike.

When JuSTEMice and CDAs join forces, a remarkable synergy occurs. JuSTEMice's focus on holistic problem-solving and data-driven insights aligns perfectly with CDAs' mission to proactively de-risk family law agreements. Together, they create a powerful alliance that not only addresses existing challenges but also anticipates and prevents potential pitfalls.

Shaping the Future of Family Law 2.0 Diversity

As we peer into the future, envisioning a family law landscape where JuSTEMice and CDAs work in harmony to provide comprehensive solutions, we see a transformative potential that includes:

1.Predictive Intelligence: By harnessing data and analytics, JuSTEMice-CDAs can anticipate needs and challenges, reducing emotional turmoil and legal battles.

2. Tailored Interventions: Holistically addressing legal, emotional, psychological, and financial aspects ensures equitable outcomes for families.

3. Continuous Improvement: Feedback and data insights refine and optimize family law agreements, evolving with changing needs.

4. Accessible Resources: The Safer Private Network (SPN) offers counseling, co-parenting guidance, and legal assistance, serving as a safety net.

5. Empowering Families: Education, awareness, and support empower families to navigate complexities with confidence and resilience.

The future of family law is not a distant dream; it's a reality taking shape. With JuSTEMice and CDAs leading the way, we're poised to usher in an era of greater fairness, understanding, and harmony in family relationships.

Quantum solutions are lighting up justice—don't get left in the lightlessness!

Chapter 10

Transforming Family Law: The Peer-toPeer Family Law 2.0 #NewDeal Options

A Revolutionary, Intentionally Unincorporated Support & Encouragement

The journey towards a more just and equitable Peer-to-Peer and Decentralized Balance Family Law 2.0 system has led us to a groundbreaking revelation: the power of intentional, unincorporated support groups. These support groups, rooted in the principles of compassion, collaboration, and innovation, form the bedrock of the Peer-to-Peer and Decentralized Balance Family Law 2.0 #NewDeal Options.

The Birth of Peer-to-Peer and Decentralized Balance Family Law 2.0

Peer-to-peer and decentralized balance Family Law 2.0, abbreviated as P2P&DB Family Law 2.0, emerged as a response to the pressing need for transformative change in device borders, the bully of biases, online and social media abuse, or family law versus law double borders with inherent double unrepresentations for kids. This grassroots movement was not beholden to corporate interests or bureaucratic red tape. Instead, it was driven by the collective determination of individuals who had witnessed the shortcomings of traditional systems and were committed to crafting a better future.

The Peer-to-Peer and Decentralized Balance Family Law 2.0 Support Group

At the heart of Peer-to-Peer and Decentralized Balance Family Law 2.0 lies its vibrant and inclusive support group. This intentional, unincorporated community serves as a haven for families navigating the complexities of device borders, the bully of biases, online and social media abuse, or family law versus law double borders with inherent double unrepresentations for kids. Here's what sets this support group apart:

1. Inclusivity: Peer-to-Peer and Decentralized Balance Family Law 2.0 welcomes families from all walks of life, recognizing that the challenges of device borders, the bully of biases, online and social media abuse, or family law versus law double borders with inherent double unrepresentations for kids know no boundaries. It fosters a sense of belonging and solidarity among its members, ensuring that no one feels alone on their journey.

2. Peer Support: Within the support group, families find kindred spirits who understand their struggles intimately. Peer support is a cornerstone of Peer-to-Peer and Decentralized Balance Family Law 2. 0, enabling members to share their experiences, insights, and solutions with one another.

3. Education and Empowerment: Knowledge is power, and Peer-to-Peer and Decentralized Balance Family Law 2.0 believes in equipping its members with the information they need to make informed decisions. Through workshops, seminars, and resources, families gain a deeper understanding of device borders, the bully of biases, online and social media abuse, or family law versus law double borders with inherent double unrepresentations for kids and their rights.

4. Collective Advocacy: The support group serves as a platform for collective advocacy. Families can voice their concerns, propose reforms, and work together to influence policy changes that benefit everyone.

5. Emotional Resilience: Navigating device borders, the bully of biases, online and social media abuse, or family law versus law double borders with inherent double unrepresentations for kids can be emotionally taxing, but Peer-to-Peer and Decentralized Balance Family Law 2.0 provides a nurturing environment where members can find emotional support and resilience-building resources.

Guaranteed Uninjury Family Laws and Trade Secret CDAs

Peer-to-Peer and Decentralized Balance Family Law 2.0 doesn't stop at moral support and advocacy; it goes a step further by offering Guaranteed Uninjury Family Laws and trade secret CDAs. Here's what these innovations entail:

1. Guaranteed Uninjury Family Laws: These laws prioritize the well-being of children and parents, representing a paradigm shift in device borders, the bully of biases, online and social media abuse, or family law versus law double borders with inherent double unrepresentations for kids. Founded on fairness, equity, and protection, families opting for Guaranteed Uninjury Family Laws can rest assured that their interests will be safeguarded, reducing emotional and financial harm.

2. Trade Secret CDAs: Childproof Deal Data Analytics (CDAs) reach a new level with Trade Secret CDAs. These solutions are designed to address each family's unique needs, offering tailor-made protection and guidance. Trade Secret CDAs act as a safety net, preventing families from falling into Contract Deprivation Zones (CDZs) and empowering them to resolve disputes effectively.

A Glimpse of the People-to-People #NewDeals

Be a part of the #NewDeal in family law – transform your legal journey!

Peer-to-Peer and Decentralized Balance Family Law 2.0' s #NewDeal Options represent a quantum leap in addressing device borders, the bully of biases, online and social media abuse, or family law versus law double borders with inherent double unrepresentations for kids. They signal a departure from traditional adversarial models towards a collaborative, empathetic, and forward-thinking approach.

In the upcoming chapters, we'll delve into the stories of families embracing the #NewDeal, showcasing how it has transformed lives and redefined possibilities within device borders, the bully of biases, online and social media abuse, or family law versus law double borders with inherent double unrepresentations for kids.

Chapter 11

Empowering Justice: A Beacon of Support and Clarity

Signature Fillable Buffers (even during Free Fillable Bindings) Anyone Can Apply

The winds of change are sweeping through the landscape of family (injury) device law. In the realm of legal empowerment, a profound shift is underway—a movement that places empathy, support, and clarity at its core. At the forefront of this movement are three transformative platforms: an Inclusive Legal Aid Hub, an Empathy-Driven Support Center, and a Clear Pathway Toolkit. These platforms are not just reshaping how families approach their legal needs; they are beacons of hope and empowerment for those navigating the complexities of double borders and double unrepresentations.

The Vision of Justice Empowerment Platforms

In a world where challenges like divorce, disability, military deployments, and device barriers can leave individuals and families feeling unheard and unrepresented, the vision for justice empowerment platforms is clear—they must offer inclusivity, understanding, and actionable support. The Inclusive Legal Aid Hub, Empathy-Driven Support Center, and Clear Pathway Toolkit embody this vision, providing a digital sanctuary where every voice is valued and every challenge is met with compassion.

1. Inclusive Legal Aid Hub: This platform redefines legal assistance by prioritizing inclusivity and accessibility. Families facing double borders and unrepresentations can access vital legal resources without the burden of prohibitive costs. The core principle of the Inclusive Legal Aid Hub is to ensure that justice is not a privilege but a fundamental right for all.

2. Empathy-Driven Support Center: Just as a lighthouse guides ships through tumultuous waters, the Empathy-Driven Support Center offers a guiding light for those navigating legal complexities. Here, families find not only legal guidance but also emotional support, understanding, and a sense of belonging. The center's empathetic approach fosters resilience and empowerment.

3. Clear Pathway Toolkit: When facing legal challenges, clarity is essential. The Clear Pathway Toolkit equips families with the tools and resources they need to navigate legal intricacies with confidence. From customizable document templates to expert guidance, the toolkit streamlines the legal journey, ensuring that clarity and understanding prevail.

Foundational Tools for Justice Empowerment

At the heart of these platforms are tools designed to empower and uplift:

1. Empathetic Advocacy: The platforms prioritize empathetic advocacy, ensuring that every individual's story is heard and respected. Families facing divorce, disability, military deployments, or device barriers receive personalized support tailored to their unique circumstances.

2. Collaborative Solutions: Collaboration is key to effective legal empowerment. Families are encouraged to collaborate with legal experts, community advocates, and peer support networks to find holistic solutions that address their legal and emotional needs.

Empower your legal rights with knowledge that enlightens and liberates!

In a world where legal disparities can exacerbate existing challenges, these platforms act as shields, offering protection and support:

1. Transparent Processes: The platforms maintain transparent processes, ensuring that families understand every step of the legal journey. Transparent fee structures and accessible resources eliminate financial barriers, empowering families to seek justice without undue stress.

2. Expert Guidance: Legal experts within the platforms provide knowledgeable guidance and advocacy, bridging the gap between legal complexities and everyday realities. Families receive clear explanations, strategic advice, and ongoing support to navigate their legal challenges effectively.

3. Real-Time Updates: The platforms keep families informed with real-time updates on legal changes, policy developments, and resources. This proactive approach empowers families to make informed decisions and stay ahead of potential legal hurdles.

Empowering Individuals with Constitutional Rights

Crucially, these justice empowerment platforms empower individuals to exercise their Constitutional rights to redress and due process. By providing accessible legal resources, transparent processes, and expert guidance, the platforms give individuals the tools they need to navigate the legal system confidently and effectively. They champion the principle that justice should be accessible to all, regardless of background, circumstance, or representation status.

In the upcoming sections, we'll delve into the inspiring stories of individuals who have found empowerment and clarity through these justice empowerment platforms. Their journeys illuminate the transformative potential of the Inclusive Legal Aid Hub, Empathy-Driven Support Center, and Clear Pathway Toolkit, offering hope and guidance to those seeking justice in the face of adversity.

Chapter 12

Embracing Transformative Deal for New Destinies

Breaking Away from Outdated Practices

The journey toward progress often demands liberation from the shackles of the past. In the domain of family law, antiquated practices and rigid agreements have long stifled innovation and growth. However, a beacon of hope illuminates the horizon—the embrace of transformative #NewDeals.

Unveiling the Shackles of Old Deals

Ancient agreements, akin to rusty shackles, have ensnared families within antiquated legal structures. These once-sturdy agreements, once revered as bastions of stability, now serve as impediments to adaptability and advancement. Let's delve into the defining traits of these archaic practices:

Inflexibility: Old deals were inflexible monoliths, offering little leeway for families to navigate shifting circumstances. They stood as rigid barriers, often incapable of accommodating the dynamic needs of modern families.

Complexity: These agreements were labyrinthine constructs, cloaked in arcane legal terminology that alienated and perplexed many families. The convoluted nature of old deals fostered confusion and hindered effective collaboration.

Lack of Empowerment: Families found themselves disempowered within the confines of old deals, with decisions imposed upon them rather than forged through collaboration. This power disparity led to disillusionment and discontent within legal processes.

Pioneering a New Era with #NewDeals

#NewDeals herald a renaissance in family law—a seismic shift toward empowerment, transparency, and adaptability. Let's dissect the defining attributes of these transformative agreements:

Flexibility: #NewDeals prioritize nimbleness, affording families the latitude to tailor agreements to evolving circumstances. This fluidity ensures that agreements remain pertinent and efficacious in addressing the ever-changing landscape of family dynamics.

Accessibility: These agreements are crafted with lucidity and conciseness, shedding the cryptic legal vernacular that once obscured understanding. #NewDeals promote accessibility, fostering informed decision-making among all parties involved.

Empowerment: #NewDeals empower families to actively shape their legal destinies. They foster a culture of collaboration and open dialogue, culminating in agreements that resonate with the unique needs and values of each family unit.

A Glimpse into the #NewDeals Revolution

Consider the tale of the Smith family, who traversed the realm of divorce armed with a #NewDeal that championed cooperative co-parenting and shared responsibilities. This innovative approach enabled the Smiths to cultivate a harmonious post-divorce relationship, prioritizing the well-being of their children.

Similarly, David, a single father grappling with child support challenges, found solace in a #NewDeal that accommodated his financial constraints while ensuring adequate support for his child. This customized arrangement empowered David to provide for his child without succumbing to undue financial strain.

The Morales family, confronted with custody intricacies, embraced a #NewDeal anchored in transparency and mutual respect. Their collaborative agreement provided a framework for ongoing communication and cooperation, fostering a stable and nurturing environment for their children.

Embrace the change that could redefine your future – join the transfiguration!

#NewDeals symbolize liberation from outdated norms, granting families the agency to sculpt their destinies and surmount legal hurdles with clarity and assurance. As we bear witness to the transformative impact of #NewDeals through real-life anecdotes, it becomes palpable that these agreements unlock the gateway to a more equitable and adaptive family law landscape.

Chapter 13

The JuSTEMice Crusade: Unleashing Cyber-Savvy Heroes and Super Siblings

Embarking on the Adventure of Change

Welcome to the exhilarating world of JuSTEMice, where ordinary individuals become cyber-savvy heroes and super siblings! As we embark on this thrilling journey of transformation, you might find yourself wondering how you can dive into this adventure and be part of a movement that's changing the game in family law. Fear not, for in this chapter, we'll unveil the secrets to joining this epic crusade and becoming a hero of justice.

The Power of One: From Zero to Hero

You may have asked yourself, "Can one person truly make a difference?" The answer is a resounding yes! Our adventure begins with everyday heroes like Madison, who ignited change with a single spark, and Alayna, whose vision turned obstacles into opportunities. These heroes prove that even the smallest actions can have a colossal impact.

Madison's Mission: A Spark of Genius

Meet Madison, whose quest for more time with her dad, Chris, led to a brilliant idea known as "Madison's Mission." By daring to think differently and taking bold steps, Madison transformed her family's dynamics and inspired others to follow suit. She reminds us that anyone can be a hero of change.

Alayna's Innovation: A Tech Marvel

Join Alayna, whose determination to break down barriers in family communication led to the creation of "Alayna's Tech Marvel." Her inventive solution ensured that families could connect without hindrance, proving that innovation knows no bounds. Alayna's journey shows us that a vision combined with action can create wonders.

Evan and Ellie's Escapade: Sibling Superstars

Dive into the adventures of Evan and Ellie, sibling superstars who turned family challenges into opportunities for growth. Their "Evan's Escapade" unlocked the secrets of effective communication, showcasing the power of teamwork and creativity. Their escapade not only transformed their family but also inspired others to embrace change.

Spreading the JuSTEMice Magic

The magic of JuSTEMice doesn't stop with individual heroes—it's a movement that thrives on collective action and shared enthusiasm.

1. Dive into Conversations: Engage in lively discussions with friends, family, and peers about the exciting world of JuSTEMice and its potential to revolutionize family law. Share captivating stories from our adventure to spark curiosity and inspire action.

2. Be an Advocate: Join forces with organizations and initiatives driving change in family law. Your voice and passion can amplify their impact and pave the way for transformative reforms.

3. Stay Curious: Stay curious and informed about the latest advancements in family law, legal tech, and justice advocacy. Your knowledge and insights will fuel your advocacy journey.

4. Champion Change: Be a fearless advocate for progressive reforms in family law practices and policies. Use your creativity and passion to craft compelling messages, participate in advocacy campaigns, and be a catalyst for positive change.

Join the digital crusade redefining heroism – enlist today!

Our adventure has shown us that JuSTEMice, CDAs, and #NewDeals hold the key to a brighter future in family law. With flexible, transparent, and empowering systems, families can navigate challenges with confidence and resilience.

You have the power to be a hero in this story. Join the JuSTEMice crusade, embrace your inner hero, and together, let's create a future where justice prevails, and every family can thrive in a world of fairness and equity.

Chapter 14

Braveheart's Barnyard Odyssey: From Peacemaker to Global Harmony Advocate

Unleashing the CoParent Dreamscape: Remote Connections and Cyberbullying Prevention

In the bustling yet tranquil sale barn, a calf named Braveheart emerged as a beacon of change. Despite its tender age, Braveheart possessed an extraordinary talent for resolving disputes and fostering harmony among barnyard dwellers.

Braveheart's Beginnings: The Barnyard Peacemaker

Picture a serene morning at the barn, with the sun casting golden hues over the fields. Two cows engaged in a heated dispute caught Braveheart's attention. With wisdom beyond its years, Braveheart stepped in and mediated the conflict, showcasing a natural flair for justice and empathy.

Word of Braveheart's remarkable abilities quickly spread, captivating animals from far and wide. Soon, Braveheart became the go-to mediator for barnyard disputes, spreading a wave of peace and cooperation throughout the barn.

The Rise of Remote Coparent Links: A Global Phenomenon

Inspired by Braveheart's success, farmers and innovators devised Remote Coparent Links—an ingenious platform for resolving conflicts and fostering understanding among animals across different farms. Braveheart's legacy expanded beyond the barnyard, becoming a symbol of peace and fairness on a global scale.

Through Remote Coparent Links, families facing turbulent times found solace and support. Brave children became champions of resilience, guiding parents through the intricate terrain of co-parenting with grace and understanding.

Braveheart's Impact: Nurturing Connections and Cyber Wisdom

Imagine barnyard discussions evolving from territorial disputes to meaningful conversations about cyberbullying prevention, social media wisdom, and psychological insight. Inspired by Braveheart's legacy, cattle engaged in advocacy for uninjured family dynamics, creating oases of harmony amidst adversarial divides.

In the heart of the CoParents Links program at MajorLeague Family.com, families like Aaron and Alayna found new beginnings and transformed their fate. Instead of succumbing to the pressures of traditional paths, they embraced harmony, respect, and enduring connections, embodying the spirit of Braveheart's legacy.

Braveheart's spirit lives on as a testament to resilience, empathy, and the transformative power of familial bonds. His odyssey from barnyard peacemaker to global harmony advocate inspires us all to nurture connections, foster empathy, and embrace the journey towards a brighter, more harmonious future.

In the words of Braveheart, "My Old New Story," echoes the sentiment that with determination, compassion, and the right support, even the unlikeliest heroes can emerge to change the course of history.

Follow the odyssey that's making peace a global affair!

Chapter 15

The Radiance of JuSTEMice: A Beacon of Brighter Futures For All

Championing JuSTEMice

As we embark on the final leg of this transformative journey, the radiance of JuSTEMice illuminates our path forward, promising a future where justice and empathy prevail. At the heart of this movement are individuals like Lisa Johnson, Emily Sanchez, and Dr. Emma Carter, whose unwavering dedication to JuSTEMice has sparked real change and laid the foundation for a brighter tomorrow.

Lisa Johnson: The Compassionate Legal Luminary

Lisa Johnson's journey to becoming a JuSTEMice champion began with a personal experience that ignited her passion for equitable legal solutions. Growing up in a family affected by divorce, she witnessed firsthand the challenges of navigating complex custody agreements. Determined to make a difference, Lisa pursued a career in law, specializing in family law with a focus on innovative and compassionate approaches.

Throughout her career, Lisa encountered numerous cases where children's voices were drowned out by legal technicalities. Fueled by empathy and a deep understanding of the emotional toll of family disputes, she became a staunch advocate for JuSTEMice principles. Lisa's innovative strategies not only secured fair custody agreements but also inspired a wave of change in the legal community, making justice more accessible and compassionate for families in need.

Emily Sanchez: A Visionary for Vulnerable Families

Emily Sanchez's path to becoming a visionary in the JuSTEMice movement stemmed from her experiences as a social worker. Working closely with vulnerable families facing custody battles and systemic injustices, Emily witnessed the profound impact of legal disputes on children's well-being. Determined to create positive change, she delved into the intricacies of family law, seeking innovative solutions to bridge gaps of injustice.

Emily's tireless efforts led to the development of groundbreaking initiatives that prioritize healing, understanding, and resilience for families. Her compassionate approach and commitment to JuSTEMice principles have transformed the lives of countless individuals, offering hope and support in times of adversity.

Dr. Emma Carter: A Trailblazer in Legal Innovation

Dr. Emma Carter's journey as a trailblazer in legal innovation began with a deep-rooted belief in the power of technology to drive positive change. As a legal scholar, she dedicated her career to pioneering AI-driven solutions in family law, recognizing the potential to revolutionize how custody cases are handled.

Driven by a passion for fairness and equity, Dr. Carter integrated AI technology with legal frameworks, paving the way for fair and just outcomes for families across the country. Her innovative work not only streamlined legal processes but also ensured that the voices of children and parents were heard and respected in every decision.

Championing the JuSTEMice Cause

Championing JuSTEMice isn't just an aspiration; it's a vital mission that requires collective effort and commitment. Individuals like Lisa Johnson, Emily Sanchez, and Dr. Emma Carter exemplify the spirit of resilience, empathy, and innovation, paving the way for a brighter and more equitable legal landscape.

Your Role in Shaping the Future

As we conclude this journey, remember that your actions matter. You can make a difference by:

1. Advocating for Reform: Support initiatives that push for family law reform and prioritize children's well-being.

2. Sharing Knowledge: Spread awareness about JuSTEMice, CDAs, and innovative legal practices that promote fairness and empathy.

3. Supporting Innovation: Encourage the adoption of technology-driven solutions like CDAs and Family Law 2.0 platforms.

4. Being an Agent of Change: Embrace the spirit of resilience and innovation, championing justice and compassion in your community.

Don't miss the glow of tomorrow's justice—shine with us!

Our journey may be ending, but the promise of a brighter future for family law and compassionate justice continues. Together, we can build a world where justice prevails, shadows dissipate, and families thrive.

Imagine Bella the dog and Max the cat, once caught in legal turmoil, now finding security and happiness in a world guided by JuSTEMice. Their stories, like countless others, remind us of the transformative power of justice and compassion.

Let us carry the torch of JuSTEMice forward, ensuring that every family, human and furry alike, finds solace and security amidst life's complexities. Together, we can create a future where justice shines brightly, and brave hearts prevail.

Chapter 16

Discovering Bravehearts at Heart

In a world teeming with digital complexities and legal conundrums, the need for clarity, support, and unified action has never been more pressing. This visual compendium seeks to illustrate the essence of initiatives and legal landmarks that aim to bridge the gap between the ideal and the real, between the letter of the law and the spirit of justice.

Through these images and descriptions, we journey into the heart of societal vacated time-spaces-to-balanceable pillars at universal cliff's edges, from local courts to national symbols, each resonating with the steadfast pursuit of fairness, support, and communal harmony. These visual narratives and accompanying explanations delve deep into the core of critical issues, illuminating the path from entrenched challenges to transformative solutions that embody our collective aspirations for a just and equitable society.

The Problem: In the United States, a child dies every six days in custody disputes within family courts, highlighting a grim landscape where child-, victim-, and family-hood (CVF) units are entrapped in socioeconomically regressive systems. These "Traporithms" — systematic or algorithmic traps in digital and socioeconomic environments — prey on vulnerable groups, leading to cyber abuse, deep fakes, and exclusionary online environments. The escalation of cyber-medium deaths, coupled with the record number of double unrepresented children across borders, underscores the urgency for a comprehensive and inclusive approach to safeguarding CVFs in an era where digital predators are rampant.

The Solution: Freedom-makers pf peer parents leveraged the LOB [Login (or child safety file) Of Buff]system, QuadCash AI, and an array of reformative programs to emancipate CVF units from the digital and socioeconomic traps. This multidimensional approach encompasses digital innovation, family law reform, community-driven safety, and financial empowerment.

LOB (Login of Buff) Kid's in Conflict Edition: An alternative to traditional two-factor authentication methods, LOB uses a dynamic, behavior-based system to authenticate users, providing a more seamless and secure access control mechanism for sensitive information and systems.

Clean Child Support: Prior to Clean Child Support, many parents faced the rigidity of outdated child support systems that did not account for the fluid nature of financial circumstances, leading to undue stress and financial burden. This icon represents the shift toward a more adaptable and fair child support system.

Circa 2011 – Children's Original Bill of rights:
As the seeds of reform were sown with the development of Clean Child Support in 2011, the legal system began recognizing the dynamic needs of families amidst financial upheaval. This period marked the start of a more compassionate approach to child support, reflective of real-world economic fluctuations and individual circumstances. The focus was on crafting laws that mirrored the principles set out in foundational rights, particularly those addressing the inherent needs of children caught in parental conflicts. Names like Christian and

Madison began to symbolize the growing awareness and dedication to rectifying these issues, setting a precedent for further legislative evolution aimed at nurturing family integrity rather than penalizing misfortune.

Christian's Law: Before Christian's Law, familial conflicts often escalated without a framework for constructive resolution, leading to broken relationships. This law, symbolized by harmonious family figures, represents the shift towards collaborative problem-solving within families.

Circa 2014:

Following the research of Christian's Law, there was a heightened focus on reducing the emotional and psychological impacts of familial disputes. By 2014, Madison's Initiative emerged as a pivotal development, addressing the prolonged separations that exacerbated family strife. This initiative recognized the importance of time in maintaining strong familial bonds and sought to minimize the delays and disruptions that often marred the judicial and custodial processes, paving the way for more resilient family connections.

Madison's Initiative: The challenge of prolonged parental separation, known as Continuance Confusion, often led to strained family bonds. Madison's Initiative, depicted by a clock merged with a heart, symbolizes the reduction of these separations, fostering stronger family connections.

Circa 2015:

Madison's Initiative laid the groundwork for further advancements in family law gap coverages, notably leading to the introduction of Alayna's Law in 2015. This new law tackled the barriers that hindered consistent and meaningful communication between separated parents and their children. By establishing more reliable communication channels, Alayna's Law sought to preserve the emotional well-being of children amidst the complexities of parental separation, ensuring that love and reassurance remained a constant despite physical distances.

Alayna's Law: Before Alayna's Law, legal and contractual obstacles prevented consistent communication between children and their parents, causing emotional distress. The icon of a phone connected with a heart represents the breakthrough in ensuring uninterrupted parent-child communication.

Circa 2016:

The communication improvements catalyzed by Alayna's Law set the stage for addressing digital-age challenges within co-parenting dynamics. By 2016, the Safer Private Network (SPN) JuSTEMice Practices and Records became crucial as it addressed the increasingly pervasive role of social media and other digital platforms that often became arenas for parental conflict, thus threatening family harmony. This initiative aimed to fortify family bonds by safeguarding interactions and ensuring they remained conduits of support rather than conflict.

Social Media Peer CoparentLinks – Social Coparent JuSTEMice Practices: The rise of social media introduced new challenges in coparenting, with digital platforms becoming battlegrounds for parental disputes, disrupting family bonds. This initiative's logo reflects the move towards protecting these bonds in the digital age.

Circa 2017:

As digital platforms began to incorporate fairer practices through the

SPN initiative, there remained a need for equity in the representation within the judicial system itself. Corey's Act, introduced in 2017, addressed these documentary and procedural imbalances that frequently disadvantaged one party over another. By promoting transparency and fairness, Corey's Act aimed to rebuild trust in the legal processes surrounding family law, ensuring all parties had a fair chance to present their cases.

Corey's Act: Previously, documentary imbalances in family law often led to unfair representations and outcomes. Corey's Act, with its balanced scale and film reels, symbolizes the journey towards equitable representation and transparency in family law.

Corca 2017:

The equitable frameworks established by Corey's Act paved the way for further protective measures in family law, exemplified by Evan's Initiative later in 2017. This initiative tackled the unilateral restrictions one parent could impose on the other, which severely disrupted family continuity and child welfare. Evan's Initiative aimed to protect children's rights to maintain balanced relationships with both parents, ensuring judicial decisions fostered their best interests rather than deepening familial rifts.

Evan's Initiative: Before this initiative, one parent could unilaterally limit the other's access to their children during legal disputes, harming family continuity. The icon showing a child figure between two adults under a gavel signifies the protection of the child's right to maintain relationships with both parents.

Circa 2018:

With a more balanced approach to family law gap, Family Law 2.0 practices discovering Evan's Initiative, attention turned towards the a second economic strains imposed on families by the child support system. In 2018, the Child Support Relief Fund was introduced to ease the financial burdens often exacerbated by processing fees, representing a shift towards a more empathetic financial structure that recognized the realities of disadvantaged parents striving to support their children amidst economic challenges.

Child Support Relief Fund: The financial burden of processing fees for child support often places undue strain on disadvantaged parents. The image of a piggy bank shielded by a courtroom column represents the alleviation of this financial strain.

Following the economic alleviations provided by the Child Support Relief Fund, 2019 marked the introduction of the RMX Innocent Parent Family Law Representative Retainer Fee Fund. This initiative addressed the critical lack of accessible legal representation, which often resulted in biased and unjust outcomes in family law cases. By offering legal defense support, this fund aimed to protect familial stability and ensure justice was attainable for all parties, particularly those financially disadvantaged.

RESCUE
MARKET X

majorleaguefamily.com

RMX Innocent Parent Family Law Representative Retainer Fee Fund: In the family court system, the lack of guaranteed legal representation for those who couldn't afford it often led to unjust outcomes, especially when homes and children were at stake. This fund's icon, featuring a protective umbrella over a family, signifies the provision of legal defense support.

As 2019 witnessed the establishment of the RMX Innocent Parent Family Law Representative Retainer Fee Fund, focusing on the critical need for equitable legal representation, the stage was set for challenging the broader legal framework itself. In the heart of Macon County, Illinois, this call for justice and integrity in family law soon echoed into a more profound critique of the legal education system. Aaron

W. Wemple's civil complaint against the University of Illinois arose from these evolving discussions, questioning the very essence and delivery of legal education. His assertion that the law should embody the clarity and immutability of scientific principles symbolized a pivotal shift, from not only securing legal aid for the disadvantaged but also demanding a foundational transformation in how legal professionals are trained. Wemple's challenge highlighted a growing discontent with a system perceived to prioritize leverage over law, setting a historical precedent for advocating a purer, more principled approach to legal education and practice.

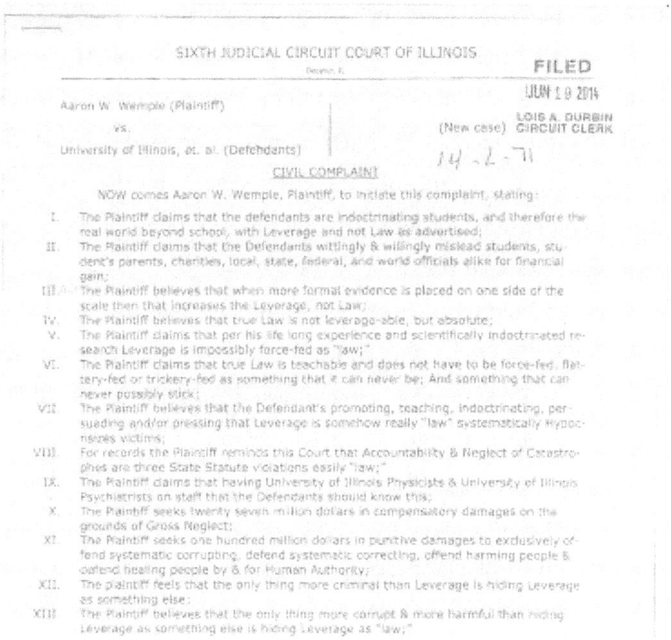

SIXTH JUDICIAL CIRCUIT COURT OF ILLINOIS

FILED
JUN 1 9 2014
LOIS A. DURBIN
CIRCUIT CLERK

Aaron W. Wemple (Plaintiff)

vs.

University of Illinois, et. al. (Defendants)

(New case)

14-L-71

CIVIL COMPLAINT

NOW comes Aaron W. Wemple, Plaintiff, to initiate this complaint, stating:

I. The Plaintiff claims that the defendants are indoctrinating students, and therefore the real world beyond school, with Leverage and not Law as advertised;
II. The Plaintiff claims that the Defendants wittingly & willingly mislead students, student's parents, charities, local, state, federal, and world officials alike for financial gain;
III. The Plaintiff believes that when more formal evidence is placed on one side of the scale then that increases the Leverage, not Law;
IV. The Plaintiff believes that true Law is not leverage-able, but absolute;
V. The Plaintiff claims that per his life long experience and scientifically indoctrinated research Leverage is impossibly force-fed as "law;"
VI. The Plaintiff claims that true Law is teachable and does not have to be force-fed, flattery-fed or trickery-fed as something that it can never be; And something that can never possibly stick;
VII. The Plaintiff believes that the Defendant's promoting, teaching, indoctrinating, persuading and/or pressing that Leverage is somehow really "law" systematically hypocrisizes victims;
VIII. For records the Plaintiff reminds this Court that Accountability & Neglect of Catastrophes are three State Statute violations easily "law;"
IX. The Plaintiff claims that having University of Illinois Physicists & University of Illinois Psychiatrists on staff that the Defendants should know this;
X. The Plaintiff seeks twenty seven million dollars in compensatory damages on the grounds of Gross Neglect;
XI. The Plaintiff seeks one hundred million dollars in punitive damages to exclusively offend systematic corrupting, defend systematic correcting, offend harming people & defend healing people by & for Human Authority;
XII. The plaintiff feels that the only thing more criminal than Leverage is hiding Leverage as something else;
XIII. The Plaintiff believes that the only thing more corrupt & more harmful than hiding Leverage as something else is hiding Leverage as "law;"

In the heart of Macon County, Illinois, a legal battle unfolded that would question the very foundations of legal education and its impact on society. Aaron W. Wemple, in his civil complaint against the University of Illinois, courageously stepped forward as a modern-day

Braveheart, challenging the giant educational institution on its own turf. His claim, rooted in the belief that law should be as immutable and clear as scientific fact, aimed to expose what he saw as a systemic flaw in the legal educational system.

Wemple argued that the institution was blurring the lines between leverage and law, thereby indoctrinating students into a system where legal outcomes could be manipulated for gain rather than grounded in absolute truth. This Braveheart of the legal realm contended that such an approach not only misrepresented the nature of law but also risked the futures of those it purported to educate and serve, turning them into potential victims of a misguided system.

He asserted that true law, like the unassailable laws of physics, should not be subject to negotiation or influence. In his view, the legal education offered was akin to a trick, a force-fed doctrine that misrepresented the essence of law, thereby endangering the integrity of the legal profession and, more critically, the societal fabric dependent on its fairness and justice.

Wemple's bold stance in the courtroom was not just a legal challenge; it was a call to arms for those who believe in the sanctity of law and education. He stood as a braveheart, demanding accountability and change, advocating for a legal education that upholds law as an absolute, not a leverageable concept. Through his fight, he sought to ensure that future generations would inherit a system that respects the truth and unyielding nature of genuine law, free from the distortions of financial or strategic leverage.

Practicing More for Families of the future:

The transformational journey from the early 2010s through to the present reflects a profound shift in family law chasm and support systems, driven by the urgent need to safeguard vulnerable children and ensure family stability. Initiatives like the "1st to 300% Race" and "1st to 600% Visitation 'Un-Trade-Ups'" represent a revolutionary approach to tackling the long-standing challenges of financial disputes and visitation rights in family law. These innovative races aim to dramatically improve financial circumstances and strengthen family bonds, showcasing a commitment to transforming the adversities of today into the secured family unity of tomorrow. Through these efforts, we see a powerful movement towards a more compassionate and equitable system that values the welfare of children and the fundamental rights of families.

Building upon the groundbreaking initiatives of the "1st to 300% Race" and "1st to 600% Visitation 'Un-Trade-Ups,'" the landscape of family law continues to evolve towards more personalized and dynamic solutions. These efforts reflect a growing recognition of the unique circumstances and needs of each family, challenging the traditional one-size-fits-all approach that often perpetuated conflict rather than resolution. This shift signifies a movement towards a more tailored and responsive legal system, poised to address the complexities inherent in modern familial structures. It sets the stage for the next conceptual breakthrough in legal reform, termed "Reflective Reversal," which aims to invert historical norms by embedding flexibility and empathy into the very fabric of legal and social frameworks, thus addressing the core problem of inflexible and uniform policies that fail to meet the diverse needs of the population.

In the vibrant town of Whiskerville, where every child had a pet by their side, a revolutionary change swept through, inspired by the newly introduced "Flexible Families Act." This act, unlike any other, acknowledged the special bond between children and their pets as essential to family harmony.

Little Timmy, an energetic eight-year-old with an adventurous spirit, had a mischievous tabby cat named Captain Whiskers. Captain Whiskers was not just a pet; he was Timmy's confidant and partner in crime. When the "Flexible Families Act" came into effect, it introduced a unique provision: Pets were now considered in family decision-making, especially in cases involving children.

On a sunny Saturday, the first meeting under the new act was held in the local community center, dubbed "Paws and Order." Children and their pets gathered, buzzing with excitement, ready to voice their needs and desires. Timmy, with Captain Whiskers perched proudly on his shoulder, stepped up to the microphone.

"Thanks to the 'Flexible Families Act,' Captain Whiskers and I get to decide on a few things," Timmy declared confidently. "For starters, we think homework should only be done if it's too rainy to play outside. And, every meal should definitely include tuna – right, Captain?"

Captain Whiskers meowed in agreement, eliciting laughter and applause from the audience.

The mayor, watching the joyous scene, nodded in approval. "This act isn't just about giving kids and pets what they want," he explained. "It's about recognizing the happiness and comfort that pets bring into our lives, making tough times a little easier."

From that day on, Whiskerville wasn't just a town; it was a place where every decision made considered the wagging tails and cheerful purrs. Homework became more flexible, tuna sales skyrocketed, and the bond between children and their pets strengthened beyond measure.

And as for Timmy and Captain Whiskers, they embarked on even greater adventures, knowing their voices mattered, cheered on by a town that valued the laughter of its children and the content purrs of their pets. Together, they proved that when families — pets included — work together, the sky's the limit!

"Reflective Reversal"

The problem statement that encapsulates the issues addressed by the above initiatives and laws we've described and experienced can be articulated as:

Problem Statement:

Historically, traditional legal and social frameworks often imposed rigid, one-size-fits-all solutions that did not accommodate the dynamic and diverse needs of individuals and families, leading to exacerbated conflicts, financial burdens, and emotional distress across various aspects of family and social law.

Common Solution:

The introduction of adaptive and empathetic legal frameworks, such as "True Followers of Jesus Bridging Heaven and Earth" and other related laws and initiatives, revolutionizes the approach by offering tailored solutions that reflect the complexities of modern life. These reforms ensure that legal and social systems are not only just but also flexible and responsive to the evolving needs of society.

This comprehensive transformation in law and policy aims to diversify traditional adversarial and punitive models with systems that prioritize healing, continuity, and mutual respect, ensuring that individuals and families receive support that is both fair and effective, mirroring the balance and fairness advocated in spiritual teachings.

Transition into Emily and Liam's Story:

In a lighter vein, consider the audacious calf, Braveheart, who, in a bold move at the auction, leaped onto a fence rather than follow the herd. This calf's unexpected stunt not only brought laughter and surprise to all present but also symbolically represents the bold leaps made by the innovative legal measures we've discussed. Just as Braveheart refused to be just another face in the crowd, these initiatives refuse to adhere to the status quo, paving new paths for individuals like Emily and Liam, whose stories of navigating family law in the digital age are up next. These narratives further illustrate the crucial impact of adaptive legal frameworks in real-life scenarios.

The Discovery of Validated Traporithms:

Drawing inspiration from the undaunted spirit of Braveheart and the moral lessons of children's tales, the concept of Traporithms was unearthed. In these narratives, the protagonists faced complex mazes of challenges and choices, akin to the digital and socioeconomic webs we navigate today. Observing these characters' journeys, researchers identified patterns of decision-making traps and coined them "Traporithms," blending "trap" and "algorithm" to describe the systematic challenges that ensnare individuals in the modern world.

A **Traporithm** is a systematic or algorithmic set of traps embedded in digital and socioeconomic environments, designed to lead individuals into predictable outcomes that often work against their best interests. These are subtle, often invisible, pathways that channel behaviors, choices, and interactions in ways that benefit the system, not the individual.

De-Risked Traporithms are those that have been analyzed and modified to minimize the negative consequences they pose to individuals. By identifying and neutralizing the harmful aspects of Traporithms, they are made safer, allowing individuals to navigate digital and socioeconomic spaces without falling prey to the detrimental effects originally designed into the system.

Shielded Traporithms refer to those that have been fortified with protective measures, ensuring that individuals can interact with these systems without being negatively impacted. This involves the implementation of safeguards and fail-safes that prevent the Traporithms from exploiting or harming users, thereby maintaining a secure and benign environment.

- Family Law Gaps Are Reformable, just like People
    - Clean Child Support: Adapts child support to changing financial circumstances.
    - Christian's Law: Promotes constructive family conflict resolution.

  - Madison's Initiative: Reduces parental separation to strengthen family bonds.
  - Alayna's Law: Ensures uninterrupted parent-child communication.
  - Social Media Peer Coparents Social Coparent Justice Initiative: Protects family bonds in the digital age.
  - Corey's Act: Advocates for fair representation in family law.
  - Evan's Initiative: Protects children's rights to family continuity during disputes.
  - Child Support Relief Fund: Eases financial burdens in child support processing.
  - RMX Innocent Parent Family Law Representative Retainer Fee Fund: Provides legal defense support for disadvantaged parents.
  - Freedomers Qualified to De-Risk and Shield Against Digital and Socioeconomic Traporithms: Specialized in safeguarding against harmful digital and socioeconomic traps.

- Community-Driven Safety
  - True Followers of Jesus Bridging Heaven and Earth:

Connects spiritual teachings with everyday actions.

- Valid Community Constructive Child Safety File Events: Enhances child safety through community collaboration.
- Valid Peer-to-Peer Family Law Gaps Support: Bridges family law support gaps through mutual aid.
- Valid Valet Walks in and Out of Clerks Offices: Streamlines legal system navigation.
- Watch in Family Court Hearing Observations: Encourages public engagement in legal transparency.
- "Un-trade-ups" in Family Law: Transforms family law into child-friendly practices.

- Financial Empowerment
  - 1st To 300% Race: Financial improvement initiative for families.
  - Mr & Mrs QuadCash 1st to 300% Race 2024 Winners: Recognizes overcoming socioeconomic challenges.
  - 1st to 300% News: Bridges traditional and modern information channels.
  - Fillable Freedoms Advocate: Promotes empowerment and support, especially for families and children.
  - Double Justice (with juSTEMice), Diversifying Billable Bindings and Fillable Freedoms: Integrates structured

and empowering systems for children and families.

True Followers of Jesus Bridging Heaven and Earth: Before the vision, individuals often felt isolated in their spiritual journeys, struggling to connect the teachings of faith with everyday actions. The image of Jesus teaching the Beatitudes brings to life the powerful connection between divine wisdom and earthly practices, offering a model for living out spiritual truths in daily existence.

Freedom-makers - Freedomeers Qualified to De-Risk and Shield Against Digital and Socioeconomic Traporithms: Freedomers are individuals or entities that are specially trained and qualified to identify, de-risk, and shield against Traporithms. They possess the expertise to dissect and understand the complex structures of digital and socioeconomic traps and implement strategies to neutralize their risks and effects. Their role is crucial in empowering individuals to navigate modern challenges safely and effectively, fostering a landscape of digital and socioeconomic freedom.

Billable Bindings Advocate: Being an advocate for Billable Bindings means championing the structured and contractual aspects of services and agreements that ensure fairness and accountability in transactions. Advocates focus on creating clear, enforceable agreements that bind parties to their commitments, safeguarding the interests of all involved, especially in contexts that impact vulnerable children and families.

Fillable Freedoms Advocate: To be an advocate for Fillable Freedoms is to support and promote opportunities that allow individuals, particularly children and families, to fulfill their potential and enjoy their rights freely. This advocacy involves working to remove barriers and create environments where people can access the resources they need to thrive, emphasizing empowerment and support at every level.

Freedomeers freedomeering through the family (injury) law versus law devices exemplify the transformative journey from systemic constraints to liberating practices. These descriptions encapsulate the essence of proactive community engagement, empathetic legal support, and innovative educational initiatives. As we delve into each aspect, from Valid Community Constructive Child Safety File Events to "Un-trade-ups" in Family Law, we uncover the collaborative efforts and strategic innovations designed to safeguard and enhance the lives of individuals and families, marking a progressive shift towards a more just and supportive legal system.

Discover the braveheart within—join the journey of heart heroes!

Embarking on Your Radiant Journey

Welcome to the grand finale of your vibrant journey! As we bid adieu to Braveheart and our inspirational champions, let's turn our focus inward and illuminate the path ahead for you.

Discovering Your Inner Light

In the timeless wisdom shared in Matthew 5:3-19, we uncover the keys to igniting your inner brilliance. Just as Lord Jesus exemplifies justice and compassion, you too can become a beacon of light and hope.

A Personal Development Odyssey: Unleashing Your Radiance

Are you ready to embark on a personal odyssey and unleash your vibrant potential? Let's walk through each stage of your transformative journey inspired by Browserbite's development process:

The Spark of Beginning: Step out of the gate with enthusiasm and curiosity. Embrace new experiences and opportunities to discover the unique light within you.

Navigating the Path: In the middle of your journey, navigate challenges with resilience and grace. Embrace setbacks as opportunities for growth and learning.

The Radiant Finale: As you approach the grand finale, shine brightly with kindness, empathy, and love. Your light has the power to inspire and uplift those around you.

Your Invitation to Radiate

As we conclude this vibrant and vital chapter for so many, we extend a heartfelt invitation to you: Embrace the light within you and let it shine for all to see. Confess "Jesus is Lord" and embark on a journey of love, compassion, and transformation. "Beleive in thine heart God hath raised him from the dead, thou shalt be saved." (Romans 10:9)

You are not just a character in a story; you are the protagonist of your own radiant tale, and future. Let's illuminate the world together with your unique brilliance and kindness. Your vibrant start, middle, and finale await, filled with endless possibilities and the joy of making a difference.

Chapter 17

Forever Brave: Inviting Jesus Into Your Heart

What does it mean to invite Jesus into your heart?

Inviting Jesus into your heart is a personal and spiritual decision to accept Him as your Lord and Savior. It involves acknowledging Jesus Christ as the Son of God, believing in His teachings, and opening your heart to His love, grace, and guidance.

How do I invite Jesus into my heart?

Inviting Jesus into your heart is a simple yet profound act of faith. You can do this through prayer, expressing your desire to accept Jesus into your life, confessing your sins, and asking for forgiveness. It's about surrendering to His will and trusting Him to lead your life.

What happens when I invite Jesus into my heart?

When you invite Jesus into your heart, you experience a spiritual transformation. You receive the gift of salvation, your sins are forgiven, and you become a child of God. This decision brings peace, joy, and a deep sense of purpose as you begin to walk in a personal relationship with Jesus.

Can anyone invite Jesus into their heart?

Yes, anyone can invite Jesus into their heart regardless of their past, background, or circumstances. Jesus's invitation is open to all who are willing to believe in Him and accept His gift of salvation. He welcomes everyone with open arms and offers redemption to all who come to Him in faith.

Is there a specific prayer to invite Jesus into my heart?

While there's no rigid formula, a heartfelt and sincere prayer expressing your desire to follow Jesus and accept Him as your Savior is powerful. You can use your own words or a simple prayer like, "Lord Jesus, I believe in You, and I invite You into my heart. Forgive my sins, lead my life, and guide me in Your ways. Thank You for Your love and grace. Amen."

What changes can I expect after inviting Jesus into my heart?

Inviting Jesus into your heart brings about transformative changes in your life. You may experience a deep sense of peace, newfound purpose, and a desire to live according to God's will. Your relationship with Jesus will grow stronger as you seek to follow Him and align your life with His teachings.

How can I continue growing in my relationship with Jesus after inviting Him into my heart?

Growing in your relationship with Jesus involves daily prayer, studying the Bible, connecting with other believers, and living out your faith through acts of love and service. Stay connected to God through regular worship, fellowship, and seeking His guidance in all areas of your life.

What if I have doubts or struggles after inviting Jesus into my heart?

It's normal to have questions, doubts, or struggles along your spiritual journey. Don't be afraid to seek guidance from mature believers, pastors, or counselors. Continue praying, studying God's Word, and trusting in His promises. Remember that Jesus is always with you, offering comfort, strength, and assurance.

Can I invite Jesus into my heart right now?

Absolutely! If you feel the desire to invite Jesus into your heart, you can do so right now. Take a moment to pray and express your faith in Him. Trust that He hears your prayers and is ready to welcome you into His loving embrace.

What is the significance of inviting Jesus into your heart?

Inviting Jesus into your heart is a life-changing decision with eternal significance. It marks the beginning of a personal relationship with God, brings forgiveness and redemption, and offers the assurance of salvation and eternal life in His presence. It's the most important decision you can ever make.

Open your heart to eternal bravery—invite transformation inside!

Chapter 18

Reflections Guide: Journeying Through "Illuminating You"

As you've journeyed through the pages of "The Law Doctor 2," you may have encountered various thoughts and questions. Here are some guiding reflections to help you navigate your reading experience from Chapter 1 to Chapter 17:

Reflection 1: Exploring Family Challenges

- What insights did you gain from Braveheart's daring escapade and the challenges faced by families like Aaron's and Lisa Johnson's?

- How did these stories resonate with your own experiences or observations about family dynamics?

Reflection 2: Innovations in Family Law

- What stood out to you about Dr. Amelia Harrison's visionary creation, Law Doctor, and the concept of Dream Open Law?

- How do these innovations in family law align with your ideas about creating a more just and compassionate legal system?

Reflection 3: Uncovering Shadowy Realities

- What shadows or challenges in family law did you discover through the stories of Madison and Chris, Lily, Alayna, Ellie, Evan, and their families?

- How do these challenges reflect broader issues within the legal system and society?

Reflection 4: Heroes and Innovative Solutions

- How did the heroic actions of Madison, Chris, Lily, Alayna, Ellie, Evan, and others inspire you to think differently about addressing family law challenges?

- What innovative solutions or approaches stood out to you as potential game-changers in the quest for a more just and inclusive family law system?

Reflection 5: Vision for Family Law 2.0

- What are your hopes and aspirations for the future of family law, considering the complexities and opportunities explored in the book?

- How can individuals, communities, and policymakers work together to create a Family Law 2.0 network that prioritizes well-being, minimizes conflict, and adapts to evolving family dynamics?

As you continue your journey through "Illuminating You," may these reflections deepen your engagement with the narrative, spark meaningful conversations, and inspire positive actions in your own life and community. Whether you're reflecting on personal growth, embracing change, or considering spiritual dimensions, may your journey through "Illuminating You" lead to meaningful insights and transformative experiences.

Journey through self-discovery that illuminates every step

Epilogue

As the final chapter of "The Law Doctor 2: Navigating the Shadows of Family Justice" draws to a close, we pause to reflect on the journey we have undertaken together. Through the highs and lows, the triumphs and challenges, we have explored the intricate landscape of family law and glimpsed the possibilities of a more just and compassionate system.

In this epilogue, we invite you to carry forward the spirit of Braveheart and the vision of Dr. Amelia Harrison. Let their stories inspire you to advocate for positive change within your communities and legal spheres. Remember that the quest for justice is an ongoing endeavor, and each of us plays a vital role in shaping the future of family law.

As we bid farewell to these pages, may the lessons learned and the stories shared resonate in your hearts, guiding you toward a brighter tomorrow for all families.

Afterword

In the afterword of "The Law Doctor 2," we extend our gratitude to all those who have contributed to this journey. From the authors and researchers to the readers and supporters, your dedication and passion have fueled the creation of this book.

As we look ahead, let us carry forward the insights gained and continue to advocate for a more inclusive and empathetic family law system. Together, we can navigate the shadows of family justice and pave the way for a future where fairness and compassion prevail.

Acknowledgements

We extend our heartfelt acknowledgments to the following individuals and organizations whose contributions have enriched the pages of "The Law Doctor 2: Navigating the Shadows of Family Justice":

- Dr. Amelia Harrison, for her visionary leadership and tireless dedication to advancing family law.

- The legal scholars and practitioners whose expertise has shaped the content of this book.

- The families who shared their stories and insights, illuminating the complexities of family dynamics.

- Our editorial team and publishing staff for their meticulous work in bringing this book to life.

- The readers and advocates who champion the cause of justice and empathy in family law.

About the author

I've walked in the shoes of Lord Jesus, Sarah, Olivia, her children, and the WAi-Family, facing countless trials without authentic remedy. Instead, I've encountered a series of legal "straight" (backstabbing) remedies, often filled with false claims and empty promises. These experiences have shaped my resolve and humility.

I am grateful for the opportunity to advocate for innovations that make technology accessible for all. From addressing digital cyber abuse to transforming family law contracts, my journey has been driven by a deep compassion and a commitment to 'un-trade-ups.' My expertise in science, technology, disability empowerment, and family dynamics is underscored by my background in Organizational Leadership, Engineering, and Psychology. This multiexperience and multidisciplinary approach helped create a fairer digital and legal landscape for everyone.

My contributions include 'The Law Doctor 2' and initiatives like Temier at www.ProgramTimier.com, which aim to enhance AI, safety, social connections, and advocacy for love and voting rights, all fortified with robust cybersecurity measures. In these challenging times, our work together offers valuable insights, guiding us through complex debates and transforming narratives for a harmonious future.

For more information about my work and contributions, please visit the *Also By* page or my Amazon.com *Author Central page*, or contact MajorLeagueFamily.com.

Also by

<u>Change the Game: One Simple New Deal Can Help Any Family Change Any Industry (Series)</u>

A Light in the Dark: Ace and Crush's Story of Courage and Determination

Adorning Your Permanent Losses So the Future You Adores It (Adaptive CLU Self-Help Guide)

Breaking the Mold: Transforming Legal Learning with Professional Law Caveats - (Family Law 2.0 Second Edition)

Dealtown Tailoring: Surviving Anguish or Success in America?

Faithfinder: Inclusive Holiday Stories Empower Readers to Find Their Superfaith!

Family Law 2.0: Hatching a Superslate (Family Law 2.0 First Edition)

Law & Side-Law: Family Hostility / Family Empathy

Inclusive Data: How Making Almighty Data Cavityism History Can Help Us Live A Better Life

IR Physics 2

Revitalizing Education: The Role of Academic Law Caveats in Student Growth

Revitalizing Inclusive Socioeconomics: The Role of SocioEcTech Justice on Human Growth

Sensory Enhancements for a More Connected World

The Law Doctor 2

Thriving Against the Odds: How to Innovate in Deprived Communities

Outside The Series:

American Pros (private)

Family Heroes Unite

Healing Hearts: The Empowering Journey of HapiHeals

I R "Freestyle" Physics – The Study of Transformation and/or Transfiguration

Lights, Camera, Color!: A Hilarious Coloring Book for Adults Inspired by 101 Laughs for Writers, Actors, Families and Friends on the Front Lines

Lights, Camera, Jokes! 101 Laughs for Writers, Actors, Families and Friends on the Front Lines

Quantum Layer Taylors Blog @ majorleaguefamily.com

The Law Doctor: With Higher Standards(Circa 2011)

Thriving Against the Odds: Kids Edition

Upbright USA (private)

Unbroken Children of God, Why Are You So Happy About Independence?

Digital Defenders: Law, Life & Pixel Protection Series:

Data Defenders: Safeguarding Your Story in a Digital World

Data Defenders Coloring Book for Adults

My Creation Paws And Claws Balance Jaws: A Coloring Book Adventure to Stay Safe in the Half-Smartphone World

Made in the USA
Monee, IL
04 November 2026

67377394R00075